INTRODUCTION TO NON-MULBERRY SILKWORMS

This book will serve as a valuable source of information on the aspects of history, current scenario, non-mulberry cultivation, pruning, pests and diseases of eri, tasar and Muga, silkworm rearing, pests and diseases of non-mulberry silkworm, processing of cocoon etc. This book can be used as resource material and practical guide for the students of agriculture, horticulture and sericulture.

Dr. D. Elumalai has devoted his academic career to Agriculture, particularly Sericulture as a Researcher and Teacher. Presently working as an Assistant Professor (Sericulture) in Department of Plant Protection at Adhiyamaan College of Agriculture and Research, Shoolagiri, affiliated to Tamil Nadu Agricultural University, Coimbatore, Tamil Nadu.

Dr. P. Mohan raj, is presently working as a Teaching Assistant in the Department of Sericulture, Forest College and Research Institute, Tamil Nadu Agricultural University, Mettupalayam, Tamil Nadu.

Dr. R. Ramamoorthy, is presently working as a Teaching Assistant in the Department of Sericulture, Forest College and Research Institute, Tamil Nadu Agricultural University, Mettupalayam, Tamil Nadu, since 2016.

Dr. C. Mohan is presently working as an Assistant professor (Agri. Entomology) in Department of Plant Protection at Adhiyamaan College of Agriculture and Research, Shoolagiri, affiliated to Tamil Nadu Agricultural University, Coimbatore. Tamil Nadu.

Dr. B. Poovizhiraja is presently working as Subject Matter Specialist (Plant Protection) in ICAR-Krishi Vigyan Kendra, Tamil Nadu Veterinary and Animal Sciences University, Chinnasalem, Tamil Nadu.

INTRODUCTION TO NON-MULBERRY SILKWORMS

Dr. D. Elumalai
Assistant professor (Sericulture)
Department of Plant Protection
Adhiyamaan College of Agricultural Research (ACAR)
Athimugam, Shoolagiri-635 105

Dr. P. Mohan raj
Teaching Assistant
Department of Sericulture
Forest College and Research Institute
Mettupalayam, Coimbatore-635 401

Dr. R. Ramamoorthy
Teaching Assistant
Department of Sericulture
Forest College and Research Institute
Mettupalayam, Coimbatore-635 401

Dr. C. Mohan
Assistant professor (Agrl. Entomology)
Department of Plant Protection
Adhiyamaan College of Agricultural Research (ACAR)
Athimugam, Shoolagiri-635 105

Dr. B. Poovizhiraja
Subject Matter Specialist (Plant Protection)
in ICAR-Krishi Vigyan Kendra,
Tamil Nadu Veterinary and Animal Sciences University,
Chinnasalem-600 51, Tamil Nadu.

CRC Press is an imprint of the
Taylor & Francis Group, an **informa** business

NARENDRA PUBLISHING HOUSE

DELHI (INDIA)

First published 2021
by CRC Press
2 Park Square, Milton Park, Abingdon, Oxon, OX14 4RN

and by CRC Press
6000 Broken Sound Parkway NW, Suite 300, Boca Raton, FL 33487-2742

CRC Press is an imprint of Informa UK Limited

British Library Cataloguing-in-Publication Data
A catalogue record for this book is available from the British Library

Library of Congress Cataloging-in-Publication Data
A catalog record has been requested

ISBN: 978-1-032-05399-8 (hbk)
ISBN: 978-1-003-19739-3 (ebk)

CONTENTS

PREFACE

The natural silks are broadly classified as mulberry and wild or non-mulberry. Non-mulberry sericulture is universally known as forest or wild sericulture. Tropical and temperature tasar, eri, muga are the principle non-mulberry silks. All branches of sericulture require food plants and manpower. In mulberry sericulture over 60% of the cost of production goes into raising and maintaining the food plants, besides a heavy initial investment is necessary for rearing houses, rearing appliances and other essentials. Likewise, among the non-mulberry varieties, eri has the disadvantage of higher production costs. Tasar is endowed by nature with vast potential. Non-mulberry sericulture is a forest-based industry uniquely suited to the economy and social structure of developing countries because of its minimum investment requirement, high employment, and foreign exchange earning potential.

This book will serve as a valuable source of information on the aspects of history, current scenario, non-mulberry cultivation, pruning, pests and diseases of eri, tasar and Muga, silkworm rearing, pests and diseases of non-mulberry silkworm, processing of cocoon etc. This book can be used as resource material and practical guide for the students of agriculture, horticulture and sericulture.

We express our gratitude to our colleagues of ACAR for their valuable support and guidance. We further extend our sincere gratitude to Mr.P. Kalamegam, Manager and Dr. H. Vijayaraghavan, Principal, Adhiyamaan College of Agriculture and Research (ACAR) for the guidance, support and motivation for preparing this book.

In the course of compilation of the text material several books, journals, technical bulletins, research papers and review articles have been consulted. I express my sincere thanks to the authors, publishers, reviewers, scientists and students whose work from the base of the book.

Dr. D. Elumalai
Dr. P. Mohan raj
Dr. R. Ramamoorthy
Dr. C. Mohan
Dr. B. Poovizhiraja

INTRODUCTION TO VANYA SERICULTURE

Introduction

Wild silks are often referred to in India as 'Vanya' silks: India produces all four varieties of natural silks viz., Mulberry silk, Tasar silk, Muga silk and Eri silk. The Tasar, Eri and Muga silk are non-mulberry silks which are wild silks and also known as Vanya Silks.

"The term 'Vanya' is of **Sanskrit origin,** meaning **untamed, wild, or forest-based.** Muga, Tasar, and Eri silkworms are not fully tamed. India **holds the monopoly on producing the Muga silk**. It is the only one cash crop in agriculture sector that gives returns **within 30 days.**

Increased importance being attached to this sector at national level is justified by the fact that it caters to the socio-economic needs of millions of tribals of rural India. It provides them with moderate earnings in the different lean seasons of the year when they do not have any work in agriculture and other allied activities. For a long time, it remained obscure as an exclusive craft of tribal and hill folks inhabiting the forests of Central India, Sub Himalayan region and NE India. In the recent past this traditional craft of tribals has assumed importance because of

- Rich production potential
- Ecofriendly nature of the activities
- Steady demand for handmade textile products within and outside the country have promoted commercial exploitation of this traditional craft into an industry of high potential. The industry has the advantage of rich natural resources such as food plants and tribal man power. The challenge is to utilize them to bring a balanced development without disturbing the existing ecological system.

History

Wild silk threads have been found and identified from **two Indus River sites, Harappa and Chanhu-daro,** dating to c. **2450-2000 BCE.** This is roughly the same period as the earliest evidence of silk use in China, which is generally thought to have had the oldest silk industry in **the world.** The specimens of threads from **Harappa appear on Scanning electron microscope** analysis to be from two **different species of silk moth, *Antheraea mylitta* and *A. assamensis*,** while the **silk from Chanhu-daro may be from a *Philosamia* species, (Eri silk),** and this silk appears to have been reeled

Wild silks were in use in China **from early times.** Moreover, the Chinese were aware of their use in the **Roman Empire and apparently** imported goods made from them by the time of the **Later Han Dynasty in the 1st to 3rd centuries CE. (Christian era).** There are significant indications in the literature that wild silks were in use in **Persia and in Greece by the late 5th century BCE, apparently referred to as "Amorgina" or "Amorgian garments" in Greece**

Vanya sericulture remained obscure for a long time as an exclusive craft of tribal and hill folks inhabiting the Central and North Eastern India. It is in the recent past that this tribal tradition assumed importance and attracted attention at National level.

The rich production potentialities within the country, steady demand for vanya silk products outside, eco-friendly nature of the production and processing activities, women participation, promoted commercial exploitation of this craft, which culminated in the transformation of this age-old tradition to an industry of immense potentiality. **Vanya silks have been commercially exploited way back in 17th Century.**

The Western World gained an appetite for these alien shaded silks in mid-1800 when a rampant silkworm disease destroyed the European sericulture industry. **Asia could not supply enough mulberry silk to cater to the needs of Europe and North America, thus creating a market for vanya silks.**

Tasar

Though there is no recorded document available regarding the origin of tasar in India, one can find the mention of tasar silk in ancient epic Ramayana "Ram's nuptial gift to Sita **includes tasar silk". Temperate tasar is of recent origin and was introduced during mid 1960's.**

Muga

The silk of Assam (Muga) was made known to the World during 1662 through a famous European traveller Jean Joseph Tavenier. Sericulture was exempted from payment of land revenue as the Kings of Assam patronised the development of sericulture. **Around 1950, there was a great earth quake in Upper Assam and the large number of muga plantations was destroyed, which hampered the growth of muga industry.**

Eri

The word eri means castor plant, is derived from the word "eranda" of Sanskrit origin. The advent of Ericulture is lost into the antiquity but, the fact remains that Assam was the original home of eri silk from time immemorial, with the earliest reference documented in 1779. The Britishers called it as "Palma Christi" silk. **The eri silk was woven into heavy clothes known as "Bar Kapoor". Captain Jenkins (1771) remarked that eri silk was of incredible durability.**

Status of vanya silk production in India

1. **Production trends:** Of the total raw silk production during 2010-11 (23,060 MT) Vanya silk contributes to around **20.58%** (4748 MT). Of the total vanya silk production, contribution of Eri, Tasar and Muga silks are **64%, 33% and 3% respectively.** Vanya silk production which was around 254 MT during 1950 gradually increased to 4748 MT during 2010-11 registering about 18-fold increase over 6 decades.
2. **Marketing of Vanya silk cocoons and yarn and products:** Though the marketing of Vanya cocoons in general and **tasar cocoons in particular is monopolized by the cocoon traders.**

   In the last two decades establishment of Raw Material Banks (RMB) in Vanya sector by CSB, Cocoon markets by Dept. of Sericulture and State government marketing agencies viz., Sericulture Federation **(SERIFED), Khadi Village Industries Commission (KVIC), Tribal Federation (TRIFED)** etc., have helped in marketing of cocoons.

   Establishment of these agencies has resulted in improvement in the bargaining power of primary cocoon producers. The Vanya silk products mainly sarees and fabrics for dress material and furnishings are being marketed mainly by manufacturers and traders by participating in various exhibitions all over India.

3. **Export of Vanya silk products:** The Vanya silk fabrics are being exported mainly from **Kolkota, Bhagalpur New Delhi, Mumbai and Bangalore** by the established exporters.

   The share of export earnings by Vanya silk products is approximately **10%** out of total export of natural silk goods.

Vanya silk promotion cell (VSPC)

Marketing is a systematic approach of understanding prospective customer and their requirement, getting the things ready, exchanging or delivering the goods/ services to the satisfaction of the customer in a profitable manner. It also includes development and maintenance of relationship with the customer which continues even after the sale is over. The products are being developed or produced keeping in view the requirement of market. Various methods are adopted by the traders to market a commodity based on its availability and demand. However, some of the commodities require special treatment so far as their marketing is concerned due to their limited base and also their clientele.

The Vanya silk products fall under this category. Muga, Tassar and Eri silkworms are not fully domesticated and we call the silks they produce as wild silks or Vanya Silk. They are basically rural produce and their handling require special care as the problems involved in the marketing of rural produce are entirely different that of urban specific and industrial products.

The marketable Vanya Silk products comprise silkworm seed, cocoons, raw silk, spun silk fabric and end products like sarees, garments, made ups etc. Out of these, fabric and other end products are the only consumer articles in strict sense.

The process of marketing is intricate and demands regular market research and intelligence to develop different sets of marketing strategies that suits the different products when producers and consumers come from different background as in case of Vanya silk products.

Chattisgarh and adjoining states like Jharkhand, Madhya Pradesh, Bihar, Orissa and West Bengal states produce Tasar cocoons.

Muga and Eri cocoons are largely produced in Assam and adjoining states of north eastern region. U.P., Himachal Pradesh, Uttarakhand and J.K. also contribute the production of Vanya silks which is not very significant.

Unlike existing marketing systems for mulberry there is no established marketing system for Vanya silk products. In the existing system, cocoon growers sell their produce to Mahajans and in turn Mahajan gets the cocoons converted into raw silk/silk yarn and then fabric through weavers.

Few states have federations or societies for more market intervention of the cocoons for level playing to control the exploitation of the cocoon growers through the Mahajans but its objectives are partially met. Various Non-governmental organizations (NGOs) are also operating in the Vanya silk producing states to provide alternate earning resource to the weaker section of the society.

Till now the slogan for development of the silk production activity was 'soil to silk' (yarn). Now, Central Silk Board has gone beyond silk reeling i.e. weaving, development of Silk products, providing linkages for sale of silk products more particularly to the Vanya Silk products by setting up a **"Vanya Silk Market Promotion Cell" (VSMPC:- 080 26282621, 26282146 &- vsmpc AT csb.gov.in)** under Catalytic Development Programme (CDP) with its headquarter in Bangalore during the year 2006-07. The activities of VSMPC are continued during the XI Plan.

Objectives of VSPC

- To provide required input support to Vanya Silk (Non – Mulberry) sector in the areas of market promotion in domestic / overseas markets.
- Design and Development of marketable products through research and development and in association with fashion designers.
- Evaluation of existing infrastructure in silk weaving clusters and need based up gradation of looms, training of weavers in advanced production techniques.
- Organizing exhibitions in major metros and non-metro cities and sponsoring manufacturers to participate in domestic and overseas marketing events.

Strategy

1. Gathering information on the producers of Vanya Silk - Organizational details, production, production capacities, product range, raw material details, present marketing arrangements etc.
2. Gathering information on domestic and export markets for Vanya silk products – the cell will engage outside experts for this purpose if necessary.
3. Collect samples of raw materials and finished products from producers in traditional production areas, Museums, Collectors, Master weavers, marketing organizations, cooperatives, NGOs, R & D Institutes etc.

4. Evaluation of samples, categorization, preparation of swatches, brochures and material for e-presentation to prospective domestic and overseas buyers.
5. Help the Industry in packaging, labeling and presentation of products – the cell will engage experts in the field.
6. Creation of e-marketing website for Vanya silks - the cell will engage experts for the purpose.
7. Facilitating participation of producers in major marketing events like fairs, expos, exhibitions etc. in the country and abroad.
8. Engage designers, merchandisers and other specialists on contract for specific periods.
9. Establishing and maintenance of Vanya Silk Shoppees in major cities.

Future Prospects or scope of the industry

- The system introduced in recent years for the organized production of **tasar block plantation**, Grainage and commercial seeds and involvement of private people and their commitment and care for scientific approach and quality have in fact induced a new zeal in tasar sector which is rewarded by way of higher production better yield and assured income to farmers.'
- Efforts taken by CSB to overcome the limitations in muga rearing such a production of good quality seeds and standardization of indoor rearing of muga silkworm.
- Efforts were taken to encourage private graineurs as that for tasar by establishment of seed production centres in isolated pockets congenial for muga rearing. (The rearers venture deep into the forest to collect natural muga cocoons from isolated patches to develop their own grainage).
- The vanya Silk showed remarkable improvement over years through various measures taken by CSB.
- On the marketing front, development of eco-friendly and market-oriented end production and exclusive marketing efforts by VSMPC (Vanya Silk Market Promotion Cell) and SMOI (Silk Mark Organization of India) has created good and ready market for Vanya silk.
- A project called "Applications of Remote Sensing and GIS in Sericulture Development" during the XI Plan. It would identify potential areas for development of silkworm food plants i across the country and develop silk such as mulberry, eri, muga and tasar through a network called Sericulture Information Linkages and Knowledge System (SILKS).

- **SILKS is a single** window information and advisory service system for the farmers providing computerized information, storage, value addition and supply of sericulture knowledge to the users in local languages.
- There is tremendous scope for improving the production and quality of silk through identification of additional potential areas and employing improved method of information collection, processing and dissemination.

DISTRIBUTION OF NON-MULBERRY SILKWORMS IN INDIA

Silks of India

- There are five major types of silk of commercial importance, obtained from different species of silkworms which in turn feed on a number of food plants.
- Except mulberry, other varieties of silks are generally termed as non-mulberry silks.
- The non-mulberry silk is recently christened as "Vanya Silk", due to its wild nature.
- **India has the unique distinction of producing all the commercial varieties of silk.**

A. Mulberry silk

Bulk of the commercial silk produced in the world comes from this variety and often refers to mulberry silk. Mulberry silk comes from the silkworm, *Bombyx mori* L. which solely feeds on the leaves of mulberry plant. These silkworms are completely domesticated and reared indoors.

The mulberry sector continues to be predominantly rural and small farmer-based, with post cocoon activities in the cottage and small industry sector. Mulberry silk contributes to around80% of the silk production.

In India, the major mulberry silk producing States are Karnataka, Andhra Pradesh, West Bengal, Tamil Nadu and Jammu & Kashmir which together contributes97% of country's total mulberry raw silk production.

B. Tasar

- Tropical Tasar growing area forms a distinct belt of humid and dense forest sprawling over the Central and Southern plateau, covering the traditional states of Bihar, Jharkhand, Madhya Pradesh, Chhattisgarh, Orissa and touching the fringes of West Bengal, Andhra Pradesh, Uttar Pradesh and Maharashtra.
- Temperate tasar (oak tasar) extends from the sub-Himalayan region of Jammu and Kashmir in the West to Manipur in the East covering Himachal Pradesh, Uttarkhand, Assam, Mizoram, Arunachal Pradesh and Nagaland.
- **Tasar (Tussah) is copperish beige colour, coarse silk mainly used for furnishings and interiors. Itis less lustrous than mulberry silk,** but has its own feel and appeal. Tasar silk is generated by the silkworm, *Antheraea mylitta D* which mainly thrive on the food plants of **Asan and Arjun.**
- There a ring is conducted outdoor in nature on the trees. In India, tasar silk is mainly produced in the States of Jharkhand, Chhattisgarh and Odissa, besides Madhya Pradesh, Maharashtra, Bihar, West Bengal and Andhra Pradesh. Tasar culture is the main stay for many tribal communities in India.

C. Oak Tasar

It is a finer variety of tasar generated by the silkworm, *Antheraea proyeli* J. in India which feed on natural food plants of oak (*Quercus*), found in abundance in the Sub-Himalayan belt of India covering the States of Himachal Pradesh, Jammu & Kashmir, Uttarakhand, Assam, Mizoram and Manipur. **China is the major producer of oak tasar in the world and this comes from silkworm *Antheraea perni.***

D. Eri

- Eri culture was mostly confined to the Brahmaputra valley of Assam in the tribal inhabited districts, followed by Meghalaya, Nagaland, Mizoram, Manipur and Arunachal Pradesh. Ericulture is introduced on a pilot scale in States like Andhra Pradesh, Tamil Nadu, West Bengal, Bihar, Chhattisgarh, Madhya Pradesh, Orissa etc.
- **Also known as Endi or Errandi, Eri is a multivoltine silk spun from open-ended cocoons, unlike other varieties of silk.** Eri silk is the product of the semi domesticated silkworm, *Philosamia ricini* that feeds mainly on castor leaves.

- Eri Silkworm being polyfagous has wide range of foodplants such as Tapioca/cassava, Papaya, Payam, Kessaru and Barkessuru etc. Eri-culture is a household activity practiced mainly in North Eastern Region for protein rich pupae, a delicacy for the tribals in the region. **Resultantly, the eri cocoons are open-mouthed and are spun.**
- The silk was used indigenously for preparation **of *chaddars* (wraps)** for own use by the tribals. Eri silk fabric is a boon for those who practice absolute non-violence and do not use any product obtained by killing any living creature.
- **Eri silk now popularized as "Ahinsa Silk". Now Eri silk is getting popular the world over due to the isothermal properties which make it suitable for shawls, jackets and blankets.** In India, Eri culture is practiced mainly in the North-Eastern States. It is also getting popularized in Bihar, West Bengal, Odisa, Uttar Pradesh and Andhra Pradesh. **Eri silk is suitable for knit products, under wears, kids wear, denim and other fashion garments.**

E. Muga

- Assam accounts for more than **95%** of the muga silk production.
- The culture is also spread in different districts neighboring Assam in Meghalaya, Nagaland, Manipur, Mizoram, Arunachal Pradesh and West Bengal.
- **This golden yellow colour silk is prerogative of India and the pride of Assam State.** It is obtained from the wild multivoltine silkworm, *Antheraea assamensis*. These silkworms feed on the **aromatic leaves of Som and Soalu plants** and are reared outdoor on trees similar to that of tasar.
- This fabric is one of the world treasures of **fine silk fabrics, woven on foot-powered, hand operated looms, which creates a subtle unevenness.** The natural shimmery golden colour of this rare, **wild silk needs no dye to enhance its exquisite beauty.**
- It is a high value product used in products like sarees, mekhalas, chaddars, etc. Muga culture is specific to the State of Assam andan integral part of the tradition and culture of that State. However, the muga culture is getting popularized to other States like West Bengal, Meghalaya and Nagaland due to the availability of Som and Soalu plants. Muga is now used to replace *zari*in sarees and for surface ornamentationis garments / apparels, etc.

VOLTINISM IN NON-MULBERRY SILKWORMS – DIAPAUSE – METHODS TO OVERCOME

Voltinism

Voltinism in *Samia ricini*

Samia ricini is a domesticated insect exhibiting multivoltine completing 5-6 generations in a year.

Voltinism in *Antheraea assamensis*

Muga is a semi domesticated species with multivoltine completing 5-6 generations in a year.

Voltinism in *Antheraea proylei*

- Oak tasar exhibits week bivoltine.
- Depending upon the geographical and agroclimatic conditions, one or two crops can be raised in a year.
- It is semi domesticated in North Eastern region and completely domesticated in North Western region.

In North eastern region

- First crop is on *Quercus serrata* during first or second week of March.

- Second crop is an additional or subsidiary crop during summer or autumn *ie,* during September by breaking diapause through photo periodic treatment of seed cocoons and making the foliage suitable by giving appropriate pruning to the plants.
- This crop is uncertain and successful crop can be harvested when temperature is stable below 26°C.

In North Western region

- First rearing is the spring crop during first or second week of March on *Quercus incana* at lower altitudes.
- Second crop is on *Quercus semicarpifolia* during last week of May or first week of June at higher altitudes.

Voltinism in *Antheraea mylitta*

- Voltinism in *Antheraea* is regulated by the environmental factors such as temperature, RH, duration of sunshine, total rainfall, No. of rainy days and day length. The life cycle of all the varieties of tasar begins with the onset of rainy season (July - Aug).
- In *univoltine* race, the life cycle is repeated only once in a year during rainy seasons (July - Aug).
- *Bivoltine race* is cultivated during *rainy period* (July - Aug) and autumn (Sept - Oct.).

  Trivoltine repeated during

  - Rainy (July - Aug)
  - Autumn (Sept - Oct)
  - Winter (Nov. - Dec.)

  In univoltine diapause extends from last part of Aug to last Part of May

  - Bivoltine → Nov - end of May
  - Trivoltine → Jan - end of May

Voltinism of tasarmoth in different altitudes

Altitude (MSL)	Local name	Voltinism	State of cultivation	Frequency of life cycle
50 - 300	BogeiSukinda	TV or BVTV	Cultivated Semi domesticated	Three/ TwoThree
301-600	NaliaDaba	BVBV	Wild semi domesticated	Two Two
601 - 1000	Modal	UV	Wild	One

- Altitude plays a significant role in the voltinism of tasar silkworm. At low altitude, voltinism is high and it is low at higher altitudes.
- *Mylitta* is trivoltine at low altitude and bivoltine at medium altitudes.
- *Antherae apaphia* is bivoltine / trivoltine at low altitude (50 - 300 ASL) bivoltine at medium altitude (300 - 600 m ASL) and univoltine at high altitude (601-1000 m ASL).

Emergence of tasar moths

Tasar moths remain under pupation for minimum period of six months ie from January to June during extreme coldness (late winter January) and extreme hotness in summer (April and May) intercepted by spring (February and March).

Generally, all the resting pupae emerge during rainy period, June on the onset of monsoon. Temperature and humidity decide the moth emergence. 26 - 28°C temperature and 72-80% RH aid in regular emergence of moths.

Diapause and its types

The development and breeding of insect may stop in the insects temporarily during he development process. This type of phenomenon of growth interruption during the life period is generally known as hibernation/ diapause.

- The hibernation can be egg diapause, larval, pupal or adult diapause.
- Diapause is defined as the period of temporary arrest in the development of insects.
- Egg diapause is seen in *Bombyx mori* and pupal diapause is seen in tasar.

Organs involved in diapause

- The brain, *corpora allata* and sub oesophageal gland are the endocrine glands involved in this process. After the end of diapause, prothoracic gland is activated.
- Low temperature as in winter increases diapauses incidence and high temperature blocks initiation of development.
- On set of rains results in termination of diapause and leads to synchronized availability of adult insects.

Factors governing diapause

- Diapause is governed by both genes and environmental factors. Thus, silkworm carries out self-regulation on its own against different types of environmental conditions. Silkworms are more susceptible to environmental conditions during different stages of development and hence environment has a very determining role in the diapause of silkworms.
- The environmental conditions influencing diapause are temperature, light, humidity and nutrition. Among these factors, temperature is the most important determining factor.

Sub oesophageal gland and its secretion

The hormone responsible for diapause is the diapause hormone secreted by the sub oesophageal ganglion which is present on the ventral side of the oesophagous in the head region.

It is connected to the brain by circum oesophageal commissure. Secretion of sub oesophageal ganglion is high in the univoltine strains followed by bivoltine strains and it is weak in multivoltine strains.

The difference in secretory capacity of the SOG among strains with different voltinism is believed to be the result of the action of the voltinism major gene that these strains possess.

This is because SOG is the major factor that directly controls Voltinism. Thus the secretion of the hormone is decided by the genetic makeup of strains under the influence of environmental factors.

The tasar silkworm seed cocoons produced during winter undergo diapause in pupal stage and resume life cycle only after the onset of monsoon when the climate is condusive for survival and there is abundant availability of food. It is essential to preserve the seed cocoons properly during the diapause period so that the loss due to erratic or unseasonal emergence and pupal mortality are minimized.

It is endeavoured to maintain the seed cocoons in such a way that moth emergence and mating begin in the second fortnight of June. The period of emergence of moth are controlled by the endogenous circadian rhythm which is affected by exogenous factors such as temperature, humidity, day length.

Temperature conditions for preservation of cocoons

Temperature should be below 35^0C & RH should not exceed 70 %. The ideal condition of RH is 40-60 %.

Due to severity of climate, there may be loss of 30-50% stock, which could be pupal mortality, erratic and unseasonal emergence. Normally winter preservation does not experience any loss due to pupal diapause. The temperature and RH are maintained by providing cooling measures such as,

- The verandah of grainage covered with bamboo mats, hessian cloth curtains or straw pads.
- Doors and windows should be covered with wet khus mats. Water is sprinkled as and when required.
- Water may be dripped on hessian cloth covered on roof Coolers designed by CTR & TI may be fixed to windows from outside.

Exposure to LD 10:14, 14:10 and 18:6 h provoked early diapause termination in *Antheraea mylitta*

Effect of latitude and altitude on diapauses in tasar, *Antheraea mylitta* Drury

Altitude, latitude, minimum temperature, maximum temperature, photoperiod and larval duration contributed to 82.60 percent for enhancing the diapause percent. Minimum temperature plays a determining role for programming this species to enter into diapause (facultative diapause).

At low latitudes, photoperiodic variation during a year is low and the stock behaves as TV. Total photoperiod received during larval span is lower in stocks behaving as TV and higher in stocks behaving as BV. Lower peak of photoperiod during yearly cycle is reflective of higher temperature and multivoltine at low latitudes. Higher altitude has similar effect to that of higher latitude, i.e., low temperature and lower number of life cycles in a year.

Termination of diapause or methods to overcome diapause

The pupal diapause of the Indian Tasar silkworm is sustained when pupae are stored in continuous darkness or exposed each day for up to 12 hr of illumination at 26°C. Diapause is terminated when pupae, one or more weeks after pupation, are exposed to continuous light or to 18 hr of daily illumination.

Termination of diapause is also brought out by brief exposure to high temperatures (10 min at 40 or 50°C) or by preliminary chilling of pupa at 2 or 5°C for 1 to 6 weeks

In oak tasar, in north eastern region, second crop is an additional or subsidiary crop during summer or autumn during September by breaking diapause through photo periodic treatment of seed cocoons and making the foliage suitable by giving appropriate pruning to the plants. This crop is uncertain and successful crop can be harvested when temperature is stable below 26°C.

TYPES OF NON-MULBERRY SILKWORMS

Types of non-mulberry silkworms of commercial importance.

Tasar silkworm

Tasar worms of **multivoltine** in nature are reared in the tropical and temperate zones. Four species of the genes ***Antheraea* Hubner** are used for commercial production. They are

Antheraea mylitta D	**- Tropical tasar silkworm (India)**
***Antheraea proylei J* (India)**	**- Temperate silkworm**
***Antheraea pernyi* G.M**	**- Chinese tasar (China and USSR)**
***Antheraea yamamai* G.M**	**- Japanese tasar (Japan)**

- *Antheraea* comprises of many species than any other genus of sericigenous insects.
- **35 species have been recorded.**
- 31 species is of Indo Australian bio graphic region and three in paleartic region, one in U.S.A. There are more than fifty variants , aberrant or races.

Muga silkworm – *Antheraea assamensis*

- The golden yellow muga silk is secreted by a **semi domesticated multivoltine species called** Muga silkworm.
- This species is widely distributed and cultured in **Assam.**
- It is **endemic** to Assam only and reared mostly in Brahmaputra valley.

Eri silkworm- *Samia ricini.*

- The brick red or white coloured eri silk produced by *Samia ricini.*
- It is a domesticated multivoltine silkworm.
- It is widely grown in Assam and in Bihar, West Bengal, Manipur, Orissa, Tripura, Andhra Pradesh and some parts of Tamil Nadu.
- It has only one species and 16 forms, variants, aberrant or races.

Systematic position of tasar, eri and Muga upto family level

Kingdom	:	Animalia
Phylum	:	Arthropoda
Sub phylum	:	Mandibulata
Class	:	Insecta
Sub class	:	Pterygota
Division	:	Endo Pterygota
Order	:	Lepidoptera
Sub order	:	Ditrysia
Super family	:	Bombycoidea
Family	:	Saturniidae

Other silk producing arthropods

- Wild silk farming is a supplementary activity for income generation for rural communities that mainly depend on subsistence farming and assist in conserving the wild silk moth and its habitat.
- Wild silk production is a unique eco-friendly agro-practice with the potential for environmental amelioration, employment and income generation.
- The present utilization of wild silk moths hardly accounts for 5% of the rich potential and most of the production is from the Far East countries.
- The steadily growing demand for silk in all consuming countries provides excellent opportunities for any country to venture into wild silk production.
- In East Africa, wild silk production would be ideal for generation of supplementary income to resource-poor farmers, reducing the destruction of their host plants, promoting conservation of the silk moths and at the same time permitting positive utilization of these biology resources by the local community.

- In Kenya, Uganda and Tanzania, 58 wild silk moth species of three lepidoptera families, Saturniidae,Lasiocampidae and Thaumetopoeidae were recorded
- A periodic survey conducted at Nagaland during 2004-2006 revealed the presence of **14 species belonging to 8 genera i.e. *Antheraea, Actias, Attacus, Archaeo attacus, Cricula, Loepa, Samia, Sonthon naxia* and a large number of host plants.**
- However, only four species are commercially exploited in Nagaland at present and there remains a great scope for producing novel silk from *Actias selene, Antherae aroylei, Samia canningi and Criculatri fenestrata.*
- The egg, worm, cocoon and adult stages of certain species have been studied for character evaluation and categorization.

Anaphe silk

This type of silkworms is seen in **southern and central Africa.** The anaphe silk is produced by worms of genus, ***Anaphe*** **which** includes

- ***Anaph emoloneyi*** **Druce**
- ***Anaphe panda*** **Boisduval**
- ***Anaph ereticulata*** **walker**
- ***Anaphe ambrizia*** **Butler**
- ***Anaphe carteri*** **Walsingham**
- ***Anaphe venata*** **Butler**
- ***Anaphe infracta***

a. ***panda*** **is most promising** for wild silk production because of huge size of cocoon and abundance of host plant, *Bridelia micrantha* Walsingham, cocoon weight: 30g – 40g.The worm spin cocoons in communes, all enclosed by a thin layer of silk.

b. It passes through seven larval instars. The developmental period took between 83 to 86 days in the dry season and 112 to118 days in the rainy season. The pupal period ranged between **158 and 178** days in the rainy season and, on the other hand, between **107 and 138 days** in the dry season. The tribal people collect the cocoons from forest and spin the **fluff into raw silk.** The **silk is soft and fairly lustrous.**

The silk obtained from *A. infracta*is known as boko and the silk from *A. moloneyi* as tissnian- tsamia and ko ko. The fabric is **elastic** and stronger than mulberry silk. Anaphe silk is used in **velvet and plush.**

Weight depends on the number of silkworms weaving the silk nest of brown colour. There are two types of nests, a medium sized nest containing **5105 worms** and a small nest containing about **20 worms.** The average weight of larvae at pre-pupal stage is **1.089g.** The moths are beautiful with cream coloured wings transverse by black lines.

Fagara silk

Fagara silk is obtained from the **giant silk moth, *Attacus atlas* L.** and a few other related species of races inhabiting the Indo Australian biogeographic region, China and Sudan. It is the largest of the living insect reaching upto **eleven inches** in wingspan. They spin light brown cocoons nearly of 6cm long with peduncles of varying lengths (2-10cm).

Host plants included *Cinnamomum, Citrus, Salix, Annona, Clerodendrum and Mussaenda*

- The natural incidence of the saturniid wild silk moth, ***Attacus atlas*** L. recorded feeding on "Tree of Heaven, *Ailanthus excelsa* Roxb. plantations at Jorhat, Assam, India.
- The incidence of the wild silk moth recorded throughout the year with peak incidence during **May-October.**
- The average cocoon weight was 14.11gand 10.29g for male and female, respectively. The average shell weight of single cocoon recorded 2.04g an1.84g for female and male, respectively which is **five times more than shell weight of domesticated eri silkworm, *Samia ricini.***
- The ***Attacus* silkworm** is not commercially cultivated because of its broken strands of silk i.e. **discontinuous filament**. There is a greater scope of spinning of cocoon to get the yarn like eri silk. This brown, wool like silk is thought to have **greater durability**. It is high time to conserve and standardize the rearing techniques of this wild silkworm. moths available in North East India which has an economic bearing on inhabitants of the region but also indirectly helps to save forest ecosystem The conservation of forest food plants like Ailanthus, *Evodia sp., Litsea polyantha* along with exploitation of wild silk moths like *Attacus atlas* provide livelihood security to the forest dwellers.

Coan silk

- Coan silk is secreted by ***Pachypa saotus*** D. These larvae are found in the Mediterranean biogeographic region (Southern Italy, Greece Romania, Turkey etc).
- This is polyphagous insect feeding on pine**, ash, cypress, juniper and oak.**
- The cocoons spun by these worms are white coloured and measure 8.9cm x 7.6cm.
- In ancient times, this silk was used to make **crimson** dyed apparel worn by dignitaries of Rome. The output from these worms is **limited.**

ERI SILKWORM FOOD PLANTS CULTIVATION AND MANAGEMENT

Sl. No.	Common name	Botanical name
I. Primary Host plant		
1	Castor	*Ricinus communis*
2	Kesseru	*Heteropanaxfragrans*
II. Secondary Host plants		
1	Tapioca	*Manihot utilissima*
2	Papaya	*Carica papaya*
3	Barkesseru	*Ailanthus glandulosa*
4	Payam	*Evodia flaxinifoliai*

- Among the food plants, castor is much preferred host due to higher water content, ash, nitrogen percentage, acidity and crude protein.

Eri silkworm food plant cultivation and management

Northeastern region of India is rich in diversity of sericigenous insects and their host plants, out of which, four types namely Muga (*Antheraea assamensis*), Eri (*Samia ricini*), Oak tasar (*Antheraea proylei*) and Mulberry (*Bombyx mori*) silkworm have been commercially exploited for silk production.

The Eri silkworm is multivoltine and polyphagous sericin producing insects. Eri culture is a traditional agro-based small scale industry, primarily practiced to meet the partial need of warm clothing. Moreover, eri pupae are popular as delicacy among the tribal people of this region. Proximate analysis of pupa showed that it contains 55-60% protein, 25-30% lipid, 4.96% fiber, and other substances, e.g. hormones, trace elements and vitamins, thus indicating that it could be a good protein source-for various purposes (Sharma, 2010).

Eri silkworm significantly contributes to the Indian commercial silk production which is mostly confined to the Brahmaputra valley of Assam in the tribal inhabited districts. Approximately, 1.3 lakh families with plantation area of 26000 hectares are involved in ericuluture in northeastern region of this country. Annual production of eri raw silk has significantly increased. It is also practiced in few districts of the neighboring states mainly Meghalaya, Nagaland, Manipur and Arunachal Pradesh. A small quantity of Eri cultivation is also spreading recently to some other parts of the country. The north eastern region of India alone produces more than 90 % of the total amount of cut cocoons and spun silk in the country.

Eri silkworm has been classified into six different homozygous strains on the basis of body colour and subdivided into eight eco-races on the basis of their ecological distribution. Cocoon colours of eri silkworm are generally white to off-white except in Kokrajhar race, which produces brick red coloured cocoon. Moth colour is variable, but always solidly white on the dorsal surface of the abdomen. Ground colour is usually light or dark grayish brown, rarely reddish, but occasionally olive gray. Wing pattern was heavily marked whiter in ante median and post median lines.

Host plant of Eri silkworm

Ricinus communis

Heteropanax fragrans

Manihot utilissima

Ailanthus glandulosa

Package of practices for eri food plant cultivation & management

Eri culture is mostly confined to the Brahmaputra valley of Assam in the tribal inhabited districts. It is also practiced in few districts of the neighbouring states mainly Meghalaya, Nagaland, Manipur and Arunachal Pradesh. A small quantity of eri cultivation is also practiced in Bihar, West Bengal, Orissa, Jharkhand, Chhattisgarh, Uttranchal, Gujarat, Andhra Pradesh etc.. The eri silkworm is multivoltine and polyphagous in nature.

Eri host plants

Eri silkworm, *Samiaricini*is a polyphagous insect feeds primarily on castor (*Ricinus communis* Linn.). However, Kesseru (*Heteropanax fragrans Seem*) is considered as another major perennial food plant. Besides these two, eri silkworm being polyphagous feeds on several alternative host plants, viz., Payam (*Evodia fraxinifolia*), Tapioca (*Manihot esculanta*), Barkesseru (*Ailanthus excelsa*) Barpat (*A. grandis*), Gulancha (*Plumeria acutifolia*), Gamari *(Gmelina arborea*), and many more.

Propagation

Propagation technique of two major food plants, Castor and Kesseru has been evolved and recommended to the field for raising systematic eri food plantation at farmers level.

Only sexual method is employed for propagation of eri host plants. In sexual method, propagation is done through seeds. In Castor, seeds are directly sown in the field for raising the plantation while in Kesseru, seeds are initially sown in nursery bed and transplanted later to the field for raising systematic plantation.

1. Castor Cultivation Package of practices

Season and climate

The ideal period for plantation is March- April in northeast region of India. September-October is another suitable season for raising castor. Castor is a warm season crop. However, it grows under diverse climatic conditions. In winter season (during long dry spell), its leaf yield gets considerably reduced. It is highly susceptible to water logging condition and hence, drainage is a must whenever low-lying area is selected for plantation.

Soil

Flat and sloppy soil either acidic or alkaline is suitable. Water stagnation is to be avoided by adopting proper drainage system in the planting area. However, humus and sandy soil suits for luxurious growth of the plant.

Varieties

A non-bloomy red variety of castor NBR-1 is popularly utilized in northeast India for eri silkworm rearing. Besides, high oil yielding varieties like GCH-5, GCH-4, DCH-519, DCH-177, CO1, Aruna etc. are being utilized for eri silkworm rearing in non-traditional states of India.

Agronomical Practices

Tillage

Land should be ploughed 2-3 times to a depth of 20-25 cm and leveled for facilitating good root penetration and easy weeding.

Spacing

Pit system of plantation is followed for raising castor for eri silkworm rearing. For sowing of seeds, 20x25x25cm (LxBxD) size pits are to be prepared maintaining 1x1m spacing. In each pit, 1 kg FYM along with Urea 13 g, SSP 25 g and MOP 3 g are to be added as basal dose and is covered with soil.

Seed collection and treatment

Mature capsule of seeds are to be collected during sunny day. Capsules are to be sun dried well and seeds to be removed. Before sowing the seeds it is necessary to treat the matured and healthy seeds with Bavistin 2gm/kg to check seed borne diseases.

Sowing

Two seeds per pit at a depth of 2.5–3.0 cm are to be sown. Germination takes place after 7-10 days. Only one healthy seedling per pit has to be allowed for vigorous growth after germination.

Application of fertilizer

In addition to organic manure (FYM), chemical fertilizer is also important for better growth and leaf yield in castor. Chemical fertilizer NPK @ 60:40:20 kg / hectare is recommended as 1st dose of fertilizer as basal dose at pit and 2nd dose, 30 kg nitrogen/ha should be applied after attaining the age of three months.

Weeding & inter-culture

A large number of undesirable weeds absorbed nutrients and moisture of soil, resulting stunted growth of castor. Regular weeding helps for luxuriant growth of castor. Besides, Ploughing, hoeing, weeding, etc. are to be carried out timely after and before application of fertilizer for healthy growth and leaf yield.

Harvesting of crop

Four leaf harvests are being made in northeast region from a single plant in a year during May-June, July-August, September-October, November-January and February-April.

Crop Yield/Acre/year (kg)

May-June 1400

July-August 1200

September-October 1200

November-February 1000

2. Kesseru food plant cultivation

Package of practices for Kesseru food plant cultivation & management

Package of Practices

Eri host plant, Kesseru is a perennial tree. Its leaves are hard and fibrous. By feeding to early stage silkworms it is difficult for worms, in chewing the leaves as compared to castor. However, cocoons harvested from the worms fed with Kesseru are compact. Hence, it takes more time for de-gumming during spinning as compared to castor fed ones.

Season

August - September is the ideal season for plantation. It requires moderate rainfall during initial period but it can withstand even high rainfall after attaining maturity.

Soil

Kesseru grows well in acidic soil. High, flat and sloppy land is better. Water stagnation is to be prevented by adopting proper drainage system.

Nursery Technique

Selection of land

For nursery, flat and well-drained land is to be selected.

Season

February is an ideal season for raising nursery.

Preparation of beds

Land should be ploughed thoroughly and make it level. Prepare 6x2 m size beds and raise the same upto 15 cm height. Apply 6 cft FYM and equal quantity of sand per bed, mixing thoroughly with the soil.

Seed collection

At the time of ripening of Kesseru fruits, it should be covered with nylon net to protect from birds. Collect the ripened fruits during February and keep the same in water overnight. Next morning, fruits are to be rubbed with a gunny cloth to remove the pulp of the seeds. Keep seeds in water to separate sunken healthy and viable seeds and reject the floating seeds.

Seed treatment

Treat seeds with Envoi-M 45 @ 2-3 gm per kg to check fungal disease.

Seed sowing

Sow 800 seeds/bed at a spacing of (15x 10) cm. Cover the seed beds with thin layer of straw to retain moisture. Daily sprinkling of water is necessary during dry season. Remove the straw after germination of seeds.

Inter-cultural operation

Attend regular weeding at an interval of 20-30 days till the seedlings attain height of 20-25 cm.

Establishment of plantation

Preparation of land

For kesseru plantation, land preparation should be started during post monsoon period i.e. last week of July. Deep ploughing of land to a depth of 20-30 cm and leveling is recommended. For plantation, prepare 30x30x30 cm pit at a distance of 2x2m spacing (in uneven hilly area 3x3 m spacing can be practiced to facilitate inter cropping). In each pit, apply 5 kg FYM and mix thoroughly with soil.

Transplantation

Transplant 6 months old healthy seedling (25-35 cm tall) to each pit. It is more preferable to plant on a rainy day.

Cultural operation

Carry out hoeing and weeding whenever required.

Maintenance of plantation

Application of manure and fertilizer

During April apply 5 kg FYM per plant once in a year. Apply NPK @125:75:25 kg/ha in two equal split doses during April and September.

Pruning

Pruning is to be done after attaining the age of three years at a height of 1.75 m preferably during February. Subsequent pruning is to be done after the interval of 3 years.

Harvesting of crop

Three leaf harvests can be done in a year during I crop: April-May, II crop: August-September and III crop: December- February.

Crop Leaf yield/Acre/Year (kg) (in 2x2 m spacing)

April-May - 4500

August-September - 3500

December-February - 2000

TOTAL = 10,000

Diseases and Pests Management in Kesseru

Kesseru plants are less susceptible to diseases and pests attack. However, attack of Termite is found in hilly region. A beetle pest, which is nocturnal in habit, sometimes damages young leaves of Kesseru plants.

Prophylactic measures

Spray 0.2% Roger or, 0.05% Demicron or, 0.07% Nuvan mixing with 0.7% Endofil-M-45 @ 1000-1200 litre per hectare 2-3 times at an interval of 10-15 days, when diseases or, pests attack is noticed.

3. Tapioca food plant cultivation

Package of practices for Tapioca food plant cultivation & management

Introduction

Tapioca (*Manihot esculenta*Crantz) a native of Brazil in Latin America was introduced to India (Kerala) by the Portuguese in the 17th Century. Popularly known as tapioca it was popularized as a food crop by Shri VisakhamTirunal, the then Maharaja of the erstwhile Travancore. The crop has gained importance as a cheap source of carbohydrate, mainly for human consumption. It's vital in tropical agriculture is due to its drought tolerance, wide flexibility to adverse soil, nutrient and management conditions including time of harvest.

Tapioca also known as "Cassava" is cultivated mostly as rainfed crop in the uplands and agency areas where a number of sago and starch mills are existing. Its fresh tubers are used commercially in Sago industries, leaves are used in eri silkworm rearing. The dried leaves of Tapioca are rich in protein, serving as an excellent cattle feed. Of late the fresh leaves are being used for rearing Eri Silk worms.

In Tamil Nadu, cassava is grown mostly in Salem, Dharmapuri, Erode, Namakkal, Cuddalore and Kanyakumari districts. In Pondicherry state also, cassava is grown extensively.

Climate

Tapioca is a tropical crop requiring warm humid climate. It requires well distributed annual rainfall when grown under rainfed conditions. It is known for its drought tolerance and hence being grown as rainfed crop successfully.

Season June-July

Soils

Light soils *viz.*, sandy loam or red loam of laterite with pH between 4.5 to 6.6 best suited. Soils of low fertility status can also be used.

Varieties and hybrids

There are a number of varieties of cassava which are grown in different regions of the country. Most of these have been developed by the Central Tuber Crop Research Institute (CTCRI), Thiruvananthapuram. Important varieties are Co 1, Co2, Co3, H97, H 165, H 226, M4, Nidhi, Sree Prakash, Sree Harsha, SreeSahya and SreeVisakham.

The cassava varieties performing better and found suited to different states are given below (state-wise)

Andhra Pradesh	Sree Sahya, Sree Visakham, Sree Prakash, Sree Jaya, H-165, H-226
Assam	Co-1, Sree Sahya, H-165, Sree Visakham
Kerala	M4, H-97, H-165, H-226, Sree Visakham, Sree Sahya, Sree Prakash, Sree Jaya, Sree Vijaya
Karnataka	H-97, H-165, H-226, Sree Sahya, Sree Visakham
Maharashtra	Sree Visakham, Sree Sahya
Tamil Nadu	CO-1, CO-2, H-97, H-165, H-226, Sree Visakham, Sree Sahya, H-119, Sree Prakash, Sree Jaya
West Bengal	Sree Sahya, H-119
North-Eastern Region	Sree Sahya, Sree Visakham

Varieties

H-226

Semi-branching hybrid and bears medium sized tubers with light brown skin having purplish patches. The flesh is white. It possesses good quality tubers. It can be harvested 7-8 months and contains 29% starch, suitable for industries. Average yield ranges between 25-30 t/ha.

H-165

Non-branching and bears short, conical tubers with golden brown sking. It can be harvested after seven months. Contain 23-24% starch. Average yield 25-30 t/ha.

Sree Sahya

Semi branching and the tubers are light brown, white fleshed, encased in a creamish rind. The starch content is 30.0% with a yield potential of 30 t/ha. Well suited for human consumption and for sago industry.

Sree Prabha

Semi spreading and the tubers are light brown, white fleshed, encased in a creamish rind. The starch content is 29.0% with a yield potential of 35-40 t/ha. Suitable for both upland and low land conditions.

Sree Prakash

Sparsely branching, short duration (7 months) selection. The tubes are brown, stout and medium sized having white flesh, yield ranges between 20-25 t/ha. The starch content is 20.0%

Sree Jaya

Sparsely branching, short duration (7 months) selection. The tubes are brown, stout and medium sized having white flesh, yield ranges between 26-30 t/ha.

M4

Non-branching table variety. It can be harvested in 8-10 months. The tubes are brown, linear and medium sized having white flesh, yield ranges between 18- 23 t/ha.

Propagation

Tapioca is propagated through stem cuttings taken from mature healthy and disease-free plants. Around 6-8 cuttings of 20 cm can be obtained from mature stem, rejecting the tender growth at the top and thick woody portion at the base.

Preparatory Cultivation

The land should be ploughed 4-5 times to a depth of 30-35 cm. Apply FYM 12.5 t/ha, in the last ploughing along with 375 kg Super phosphate (60kg P2O5) and 50kg Lindane dust (to control termites) and incorporate in the soil by ploughing. Prepare the land into flat beds with good drainage channels.

Planting of Rooted Cuttings

The mature stems should be made into 20 cm cuttings using any sharp implement (country knife or Kattipeetha) without damaging the buds. The cuttings are to be immersed in a solution of Dithane M-45 (3 gram) + Dimethoate 2 ml/lit of water for 5 minutes and then planted in a raised nursery bed, side by side for 7-10 days with daily watering to allow them to initiate rooting. The rooted cuttings are to be planted in the main field at 90 x 90 cms to a depth of 5cm inside the soil. There should be optimum moisture at the time of planting. There should be 12,345 plants per ha. For which about 13,000 rooted cuttings are to be maintained including those for gap filling.

Planting season

Under irrigated conditions, crop is planted during December to February while it is planted in July - September as rainfed crop.

Manures and Fertilizers

A fertilizer dose of 60:60:60 kg of NPK/ha has to be applied along with FYM 12.5 t/ha. Whole P_2O_5 is to be applied as basal dose in the last ploughing. N and K_2O are to be applied in three equal split doses at 30, 60 and 90 days after planting. They are to be applied around the plant by making a ring at 10-15cm distance from the plant and cover with soil.

Fertilizer is applied in two to three split doses as basal or top applications after weeding the crop. Farmers apply fertilizers indiscriminately both under irrigated and rainfed conditions. Fertilizers are applied after 2 to 3 months of transplanting in the main field at an interval of 15 to 20 days. N, P2O5, K2O are applied @ 65: 100: 170 kg ha^{-1} under irrigated conditions and 25 : 25 : 150 kg ha-1 under rainfed conditions in Tamil Nadu.

Intercultivation

It is an important operation in Tapioca cultivation. Light digging or hoeing should be given at least thrice during early stages to remove weeds. Two healthy shoots per plant have to be retained on opposite side by removing the rest.

Intercropping

Gives an additional net income of Rs. 3000-5000/ha within 3-3 ½ months. Utilize light, water and nutrients more effectively from the interspaces of cassava. Control weeds and adds organic matter and nitrogen to the soil.

Cultivation detrails of intercrops

Name of intercrop	Cultivar	Duration (days)	Spacing (cm)	No. of rows	Seed rate (kg/ha)	Fertilizer NPK (kg/ha)	Yield (kg/ha)
Groundnut	TMV-2	100	30×20	2	40-45	10:20:20	1200
	TMV-7	100			(kernel)		
	Pollachi-2				(dry pod)		
French bean	Contender	70	30×20	2	40	20:30:40	2000
Cowpea (grain	S-488	90	30×15	2	20	10:15:10	800
Cowpea (vegetable)	B-61 (Arka garima)	65	90×20	1	8	10:15:10	3000

Planting the main crop and intercrop

Select only bushy types of intercrop which mature within 100 days. Plant cassava in the month of May-June at a spacing of 90x90 cm. Dibble the intercrop seeds immediately after planting of cassava. Basal dress cassava as per schedule given for pure stand.

Apply the recommended dose of NPK to the intercrops about 30 days after sowing followed by light interculturing. Top dress cassava immediately after harvest of intercrops with the recommended dose of fertilizers and earth up.

Irrigation

Each rooted cutting has to be given pot-watering at the time of planting if there is no adequate rain or moisture in the soil. Later on, if dry spell prevails, irrigate the crop at 15-20 days interval in chalka soils.

Intercropping

Intercrops like Green gram, Black gram, Groundnut and Maize can be grown to derive an additional income of about Rs. 1500/- per ha. within 2 ½-3 months. The intercrops have to be sown along with planting of Tapioca and fertilized separately as per their fertilizer requirement. They are to be harvested before 90 days.

The practice of growing intercrops is seldom followed in Tapioca in East Godavari District of Andhra Pradesh as it comes in the way of frequent operations of local 'gorru' in between the rows by the farmers for checking weed growth and to make the soil prous for better growth and development of tubers.

Harvest

- Tapioca becomes ready for harvest by 7-8 months after planting.
- Harvesting is done by digging with crow bars. The fresh tubers are highly perishable and can be stored only for 2-3 days.
- Tubers may be cut into chips and sun dried for a week and stored with 12-13% of moisture content for 2-3 months in air tight containers.

Cassava varieties for leaf protein-an import substitution product for pet feed and cattle feed

Cassava leaves have high protein content (14-20 per cent) though the roots are low in protein (1-2 per cent). The leaves, after processing are used for protein enrichment of different feeds and have been found cheaper than the imported Alfalfa leaf protein. This is being utilized commercially in feed industry in Thailand and other Asian countries growing cassava for commercial use.

Post-harvest utilization

The potential of cassava as poultry, pig and fish feeds as well as cassava leaf for eri silk worm rearing also.

Post-harvest Technology

The planting material is to be stored as whole stem under the shade of trees with stems in vertical position for next planting season. They should be treated with fungicide like Dithane M-45 (3 g) /) + Malathion (2 ml) or Chloripyriphos (2 ml) per litre of water to prevent the incidence of diseases and pests during storage.The planting material can also be stored safely in " Zero energy cool chanber method".

6

PESTS OF CASTOR AND THEIR MANAGEMENT

Castor is attacked by more than twenty pests of which capsule borer, hairy caterpillars, other defoliators, leaf hopper and white fly are serious.

Major pests				
1.	Capsule & Shoot Borer	*Conogethes punctiferalis*	Pyraustidae	Lepidoptera
2.	Castor semi looper	*Achaea janata*	Noctuidae	Lepidoptera
3.	Slug caterpillar	*Parasa lepida*	Cochilididae	Lepidoptera
4.	Hairy caterpillar	*Euproctis fraterna*	Lymantriidae	Lepidoptera
5.	Hairy caterpillar	*Portrhesia scintillans*	Lymantriidae	Lepidoptera
6.	Tussock caterpillar	*Notolophus posticus*	Lymantriidae	Lepidoptera
7.	Hairy caterpillar	*Dasychira mendosa*	Lymantriidae	Lepidoptera
8.	Castor butterfly/ spiny caterpillar	*Ergolis merione*	Nymphalidae	Lepidoptera
9.	Wooly bear	*Pericallia ricini*	Arctiidae	Lepidoptera
Minor pests				
10.	Leaf hopper	*Empoasca flavescens*	Cicadellidae	Hemiptera
11.	White fly	*Trialeurodes ricini*	Aleyrodidae	Hemiptera
12.	Thrips	*Retithrips syriacus*	Thripidae	Thysanoptera
13.	Castor gallfly	*Asphondylia ricini*	Cecidomyidae	Diptera

I. Borers

1. Capsule & Shoot borer: *Conogethes punctiferalis* (Pyraustidae: Lepidoptera)

Distribution and status: India, Australia, Burma, Sri Lanka, China, Indonesia and Malaysia.

Host range: Castor, mango, sorghum ears, guava, peaches, cocoa, pear, avacado, cardamom, ginger, turmeric, mulberry, pomegranate, sunflower, cotton tamarind, hollyhock.

Damage symptoms: The damage is caused by the caterpillar, which bores into the main stem of young plant and ultimately into the capsules. The borer is distributed throughout India where castor is grown.

Bionomics: Adult is medium sized with small black dots on pale yellow wings. It lays eggs on the developing capsules. Egg period is 6 days. Larva measures 24 mm when fully grown. Larva is pale green with pinkish tinge and fine hairs with dark head and prothoracic shield. Larva lives under a cover of silk, frass and excreta. Larval period is 12-16 days. It pupates in the stem or capsule.

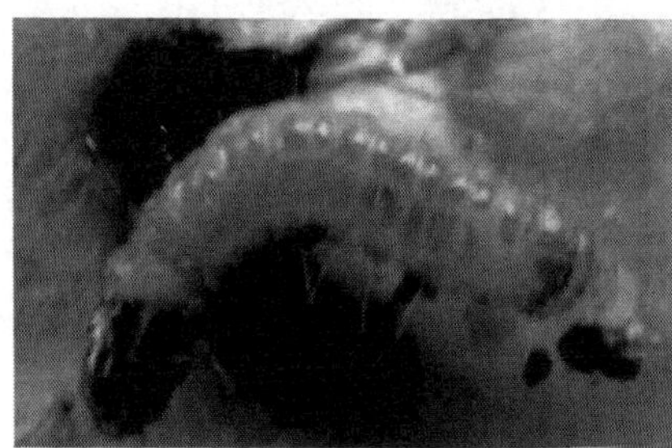

Management: Spraying the infested crop with endosulfan 35 EC 2.0 L (or) carbaryl 50 WP 2 kg or methyl parathion 50 EC 2.0 L @ 1000-1200 L water per hectare proved effective in controlling the pest.

II. Leaf feeders

2. Castor semi looper: *Achaea janata* (Noctuidae: Lepidoptera)

Distribution and status: India, Pakistan, Sri Lanka, Thailand, Laos, Malaysia, Philippines.

Host range: Castor, rose, pomegranate, tea, citrus, mango, *Cadiospermum helicacabum*

Damage symptoms: The damage is caused by both the caterpillar and adult moth. The caterpillars feed voraciously on castor leaves. Feeding from the edges inwards, leave behind only the mid rib and the stalk. The damage is maximum in August, September and October. The adult of this species are fruit sucking moths and cause serious damage to citrus crop.

Bionomics: Adult is a pale reddish-brown moth with black hind wings having a median white spot on the outer margin. Eggs are laid on the tender leaves. Egg period is 2-5 days. Larva is a semilooper with varying shades of colour with black head and a red spot on the third abdominal segment and red tubercles in the anal region. Larval period is 11-15 days. It pupates in soil for 10-14 days. (*Parallelia algira* looks very similar to *Achaea janata* but the wings have black stripes or triangles)

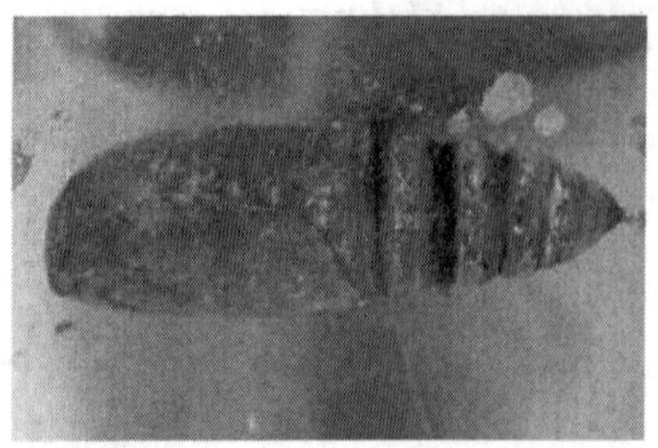

Management

1. Dusting the infested crop with 2% parathion dust @ 20-25 kg/ha.
2. Spray endosulfan 35 EC 2.0 L or carbaryl 50% WP 2 kg in 1000-1200 L water/ha.
3. Conserve braconid parasitoid *Microplitis ophiusae* since it keeps the pest under check. (Cocoons are often seen on the ventral surface of the posterior side)

3. Slug caterpillar: *Parasa lepida* (Cochilididae: Lepidoptera)

Distribution and status: India, Malaysia, Sri Lanka, South East Asia.

Host range: Castor, pomegranate, citrus, coconut, palm, rose, wood apple, country almond, mango, palmyrah, cocoa, coffee, banana, rice and tea.

Damage symptoms: Larva feeds on leaves voraciously leaving only the midrib and veins resulting in severe defoliation.

Bionomics: Adult moth is green with brown band at the base of each forewing. Eggs are laid in groups and covered with hairs on the leaves. Egg period is 4-5 days. Larva is stout, slug like ventrally flat, greenish body with white lines and four rows of spiny scoli tipped red or black; larval period is 40-45 days. It pupates in plant as cocoons covered with irritating spines and hairs

Management: Spray endosulfan 2.0 L in 1000 L of water per ha

4. Hairy caterpillar: *Euproctis fraterna* (Lymantriidae: Lepidoptera)

Distribution and status: India:

Host range: Castor, linseed, groundnut pigeonpea, grapevine, cotton, pomegranate, mango,coffee, pear and rose

Damage symptoms: Defoliation is the main symptom. The pest is active throughout the year but its activity is reduced in winter.

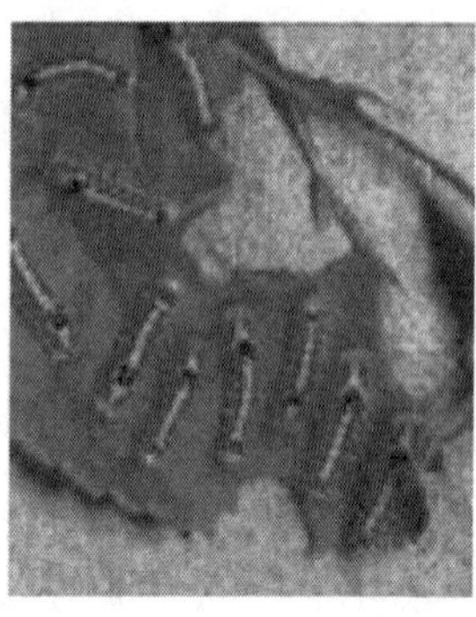

Bionomics: The adult moth is yellowish with pale transverse lines on fore wings. It lays egg in groups on lower surface of the leaves. The egg period is 4-10 days. The caterpillar possesses red head with white hairs around and a long tuft and a reddish-brown body with hairs arising on warts and a long pre- anal tuft. There are six larval instars. The larval periods last for 13-29 days. It pupates in a silken cocoon in leaf folds for 9-25 days. The larva over-winters during winter season.

Management

1. Release larval parasitoids *viz.*, *Helicospilus merdarius, H. horsefieldi, Apanteles* sp., *Disophrys* sp.
2. Dust the infested crop with parathion 2 D @ 20-25 kg per ha or malathion 5 D 25- 30 kg/ha (or) carbaryl 10 D @ 20 kg/ha.

5. Hairy caterpillar: *Porthesia scintillans* (Lymantriidae: Lepidoptera)

Distribution and status, damage symptoms and management as given for *Euproctis fraterna*

Host range: Castor, rose, cotton, redgram, mango, linseed, gogu and sunnhemp

Bionomics: Larva has yellowish brown head, a yellow dorsal stripe with a central red line on the body and tufts of black hairs dorsally on the first three abdominal segments. Adult is yellowish with spots on the edges of forewings. Life cycle is very similar to that of *Euproctis fraterna.*

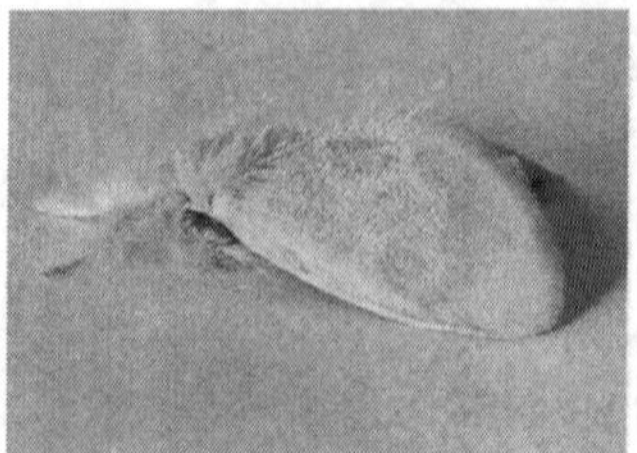

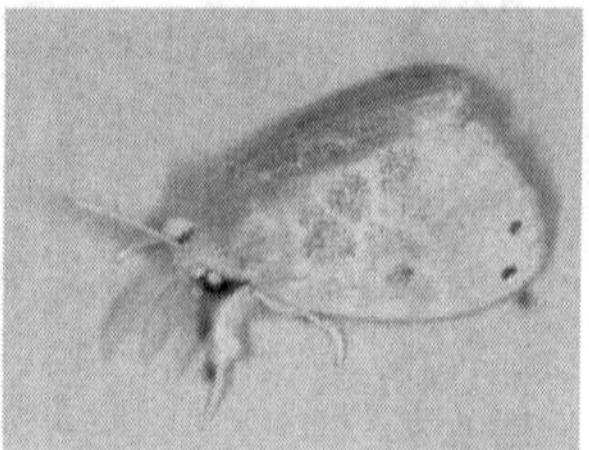

6. Tussock caterpillar: *Notolophus posticus* (Lymantriidae: Lepidoptera)

Distribution and status, damage symptoms and management as given for *Euproctis fraterna*

Host range: Castor

Bionomics: Male is winged and female being apterous, sluggish cling to the cocoon after emergence. Males are attracted to the females at dusk. Females lay 350 cream coloured subspherical eggs in mass on the cocoon itself.

Egg period 7 days and larval period 16 to 19 days. Larva has brown head with a pair of long pencils of hair pointing forward from prothorax, tuft of yellowish hairs laterally on first two abdominal segment and dorsally on first four abdominal segments and long brown hairs dorsally from 8th abdominal segment. It pupates in transparent silken cocoon inside leaf roll.

7. Hairy caterpillar: *Dasychira mendosa* (Lymantriidae: Lepidoptera)

Distribution and status, damage symptoms and management as given for *Euproctis fraterna*

Bionomics: Adult is yellowish brown moth. Larva is greyish brown with dark prothoracic and preanal tufts of hairs. Prolegs are crimson coloured.

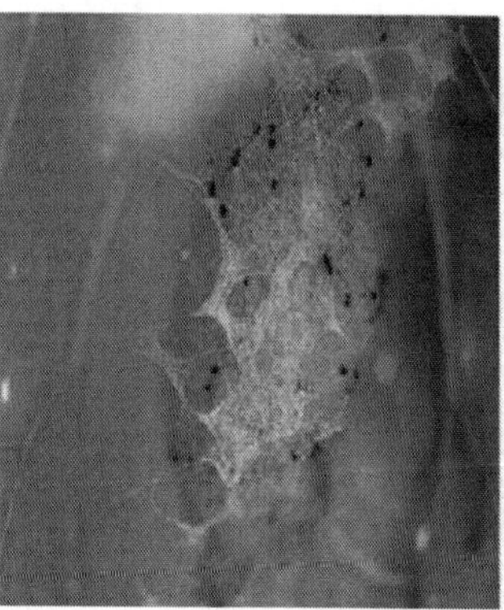

8. Castor butterfly/spiny caterpillar: *Ergolis merione* (Nymphalidae: Lepidoptera)

Damage symptoms: It is a serious though sporadic pest. Insect attacks the crop at an early stage. Insects feed on the leaf tissue and cause defoliation.

Bionomics: Brown butterfly with black wavy lines on the wings. Larva green coloured, spiny (spines branched at the tip) caterpillar with yellow stripe on the dorsal region. Pupates in a brown chrysalis.

The adult lays dome shaped, shiny white eggs singly on the underside of the leaves. Single female lays 42 to 50 eggs during her life span. The eggs hatch in about a week. The duration of the pupal stage lasts 5-6 days in September to October and 8 to 9 days in December to January. The life cycle of the pest is completed in 20 to 21 days in August to September and 37 to 42 days in December to January.

Management

1. Collect and destroy promptly the affected leaves, etc., which contain larvae inside.
2. Dust the infested crop with parathion 2 D @ 20-25 kg per ha or malathion 5 D 25 - 30 kg/ha (or) carbaryl 10 D @ 20 kg/ha.

9. Wooly bear: *Pericallia ricini* (Arctiidae: Lepidoptera)

Damage symptoms: The damage is caused by caterpillar. It feeds on leaves resulting in defoliation.

Bionomics

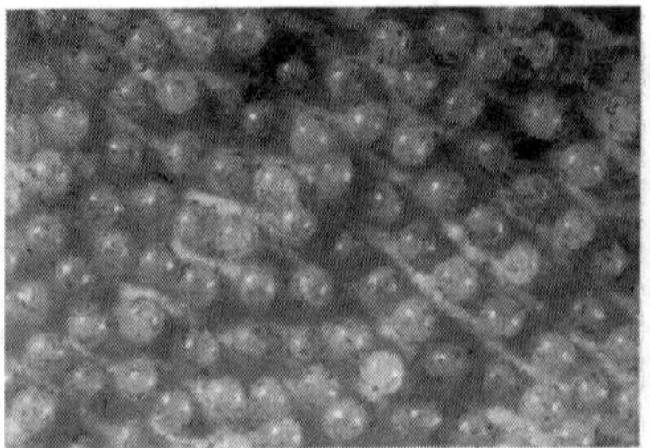

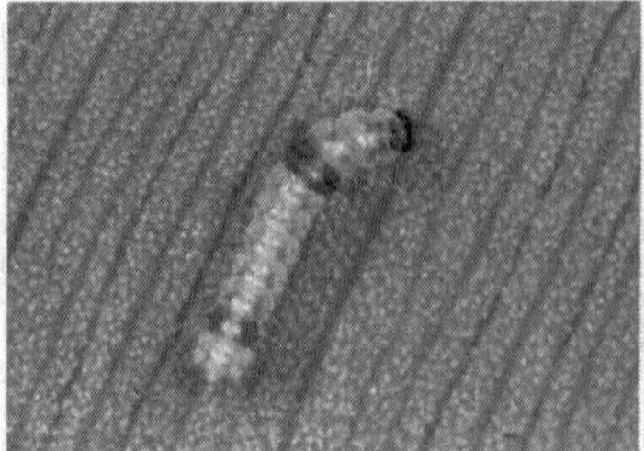

The larva is robust, greyish black or blackish brown with red head and thick tuft of hairs arising from the body. The adult is greyish brown or black with black spots on wings. Hind wings are pink or red colour with black spots.

Management

- Collect and destroy the caterpillars.
- Dust the infested crop with parathion 2 D @ 20-25 kg per ha or malathion 5 D 25 - 30 kg/ha (or) carbaryl 10 D @ 20 kg/ha for young caterpillars.
- Spray endosulfan 35 EC 2.0 L or carbaryl 50% WP 2 kg in 1000-1200 L water/ha.

10. Leaf hopper: *Empoasca flavescens* (Cicadellidae: Hemiptera)

Damage symptoms: Nymphs and adults suck the sap from the under surface of the leaves and cause "hopper burn". Leaves become crinkled and cup shaped.

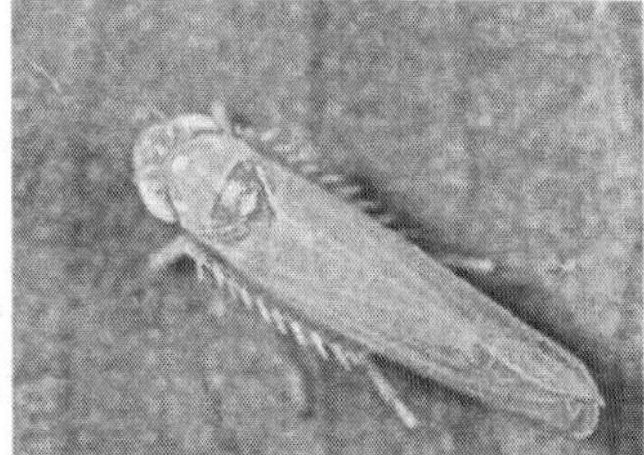

Bionomics

- Adult is green, wedge shaped hopper.
- It lays eggs within the leaf veins. A female lays 15-37 eggs during an
- oviposition period of 5-7 days. The egg period is 7-8 days. The nymphal period is 9 days.

11. White fly: *Trialeurodes ricini* (Aleyrodidae: Hemiptera)

Damage symptoms: Water soaked spots on the leaves which become yellow and dried. Colonies of whitefly are found on the under surface of leaves.

Bionomics: The adults are pale yellow with white wings covered with waxy powder. It lays eggs in clusters on the under surface of leaves. Nymphal stage undergoes four instars. The life cycle is completed in 19-21 days during July-September.

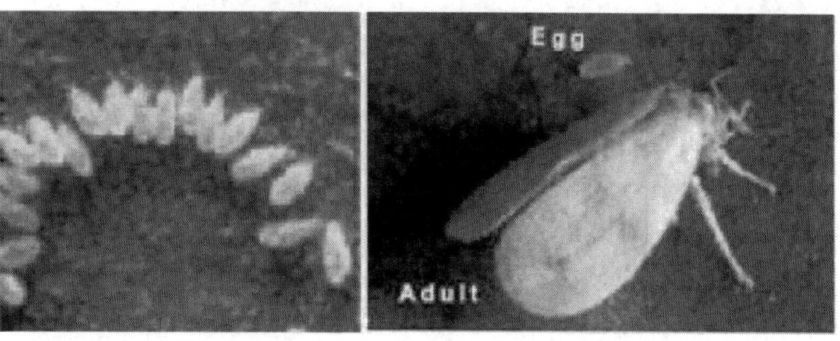

12. Thrips: *Retithrips syriacus* (Thripidae: Thysanoptera)

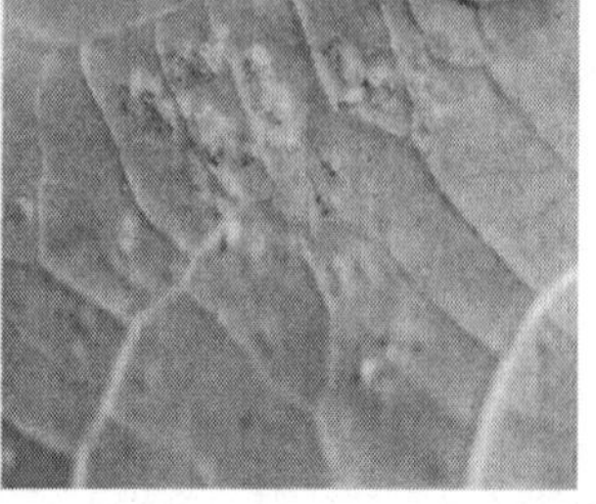

Damage symptoms: Nymphs and adults feed on the upper surface of the leaves.

Affected leaves give a shiny appearance.

It is also found on cotton and rose.

Bionomics: Pinkish nymph, black adult with fringed wings.

13. Castor gallfly: *Asphondylia ricini* (Cecidomyidae: Diptera)

Damage symptoms: The damage is caused by maggots. As a result of feeding by them, the buds develop into galls and produce no fruits and seeds. This pest is active from September to March.

Bionomics

- Adults are a mosquito like small fly. The female lays eggs singly in the buds. Incubation period is 2-4 days.
- The young maggot feeds on the floral parts and cause malformation of buds which fail to develop into seeds.
- Larval period lasts for 14-21 days.
- The pupal period is 7-12 days. Complete life cycle takes 23-37 days.

Management: Spraying the infested crop with 0.07% endosulfan 35 EC or 0.05% methyldemeton 25 EC @ 1000-1200 L water/ha.

Integrated pest management of castor crop

I. Cultural method

1. Resistant varieties: (a) Variety C3 Pakistan is tolerant. R.C.1098 and R.C.1096 coonoo are resistant to jassid attack. (b) Varieties R.C.1066, R.C.1067, R.C.1092, R.C.1069, R.C.1071 and R.C.1072 are resistant to mite infestation.
2. Summer ploughing: Deep summer ploughing should be followed, so that the larvae of semilooper, hairy caterpillar pupated in the soil will be killed due to scorching sunlight.

II. Mechanical method

1. Set up light trap to attract and kill lepidopteran moths
2. Collect and destroy the egg masses of *Spodoptera litura* and slug caterpillar.

3. Collect and destroy the early instar larvae of *Spodoptera litura*, semilopper and hairy caterpillar.

III. Microbial method

1. **Use of bacteria:** Spraying of thuricide (*Bacillus thuringiensis*1%) is found to be effective in controlling the larvae of *A. janata* and other lepidopterous larvae.
2. **Use of virus:** *Nuclear polyhedrosis, Cytoplasmic polyhedrosis* and pox-like virus has been found effective against *A. moorei* and *Euproctis* spp.
3. **Use of nematodes:** *Mermis submigrescens* have been found effective against *A.moorei.*
4. **Use of antifeedants:** Triphenyl tin compound 45% WP @ 0.06% and other fentin compounds will protect the crop from the attack of *Spodoptera mauritia, Spodoptera littoralis, Pericallia ricini, Spodoptera litura.*
5. Apply NSKE 3% + neem oil 2% for the control of semilooper.
6. Apply dimethoate 500 ml/ha or methyl demeton 25 EC 1500 ml/ha to control sucking pests.
7. Apply endosulfan 4D 25 kg/ha to control semilooper and other pests.
8. Spray any one of the following insecticides/ha thrice from flowering at three weeks interval to control capsule and shoot borer. Malathion 2 L, and carbaryl 50 WP 2 kg in 1000 L of water.

PESTS OF TAPIOCA AND THEIR MANAGEMENT

Tapioca

S. No	Common name	Scientific name	Family	Order
1	Cassava scale	*Aonidomytilus albus*	Diaspididae	Hemiptera
2	Black scale	*Parasaissetia nigra*	Coccidae	Hemiptera
3	White fly	*Bemisia tabaci*	Aleyrodidae	Hemiptera
4	Spiraling whiteflies	*Aleurodicus dispersus*	Aleyrodidae	Hemiptera
5.	Rose thrips	*Retithrips syriacus* *Rhipiphorothrips cruentatus*	Thripidae	Thysanoptera
6.	Mite	*Tetranychus urticae*	Tetranychidae	Acarina

Cassava scale: *Aonidomytilus albus* (Diaspididae: Hemiptera)

- This is a hard scale, oval and mussel-like. Male is winged. Eggs are laid inside scale. They hatch in 4 days. Nymphs are active and move on stems spreading to new areas of new stems.
- They settle close to one another, feed on sap and become full grown in 20 to 25 days. Scales infest stems.
- Leaves of attacked plants become discoloured and dry up. In severe cases desiccation of the stem and death of plants occur. Stunting of the plants results from thousands of the scales feeding on the stems. Pest is distributed through movement of crawlers and infested stems.

IPM

1. Select pest-free setts for planting
2. Collect and burn the stems infested with scales
3. Encourage activity of coccinellid predators, *Chilocorus nigritus*
4. Spray parathion 0.5% or methyl demeton 0.25% or malathion 0.1%

Whitefly: *Bemisia tabaci* (Aleyrodidae: Hemiptera)

- Nymphs and adults feed on cell sap from leaves and cause chlorotic spots and yellowing and drying of leaves.
- Nymphs are greenish yellow, oval in out line along with pupae on under surface of leaves. Adults are minute with yellow body covered with white waxy bloom.

Spiraling whiteflies: *Aleurodicus dispersus* Aleyrodidae Hemiptera

- It is an introduced polyphagous pest of vegetables, fruit trees, ornamentals and shade trees. It is native of the Caribbean Islands and Central America.
- It is widely distributed in almost all countries due to rapid dispersal and adaptability.
- It is found on 128 plants including guava, cassava, cotton, chillies, tomato, brinjal, bhendi, papaya, crotons and weeds such as *Euphorbia, Corchorus, Eclipta, Vernonia, Vicoa, Acalypha, Alternanthra, Amaranthus, Convolvulus, Abutilon* etc.

Biology

Adults are larger than whitefly species and white in colour with waxy coating on the body. Eyes are dark reddish brown. Fore wings are with three characteristic spots. Eggs are laid in a spiraling pattern (concentric circles) on the undersurface of leaves. Egg period lasts for 5-8 days. Nymphal period is 22-30 days. Adult longevity is for 13-21 days. Total life cycle is completed in 40-50 days.

Symptoms

Adults and nymphs congregate heavily on the lower surface of leaf, suck the sap and cause pre-mature leaf drop, chlorosis, yellow speckling, crinkling and curling. Honey dew secretion also leads to the development of sooty mould fungus.

The copious white, waxy flocculent material secreted by all the stages of the pest is readily spread by wind and thus cause public nuisance.

It is also a suspected vector of mycoplasma disease, lethal yellowing in coconut.

IPM

1. Remove and destroy damaged leaves along with life stages.
2. Remove and destroy weed plants like *Abutilon, Acalypha, Euphorbia,* etc., in the nearby vicinity as these plants are alternate hosts.
3. Use yellow sticky traps at 15/ha to attract and kill the adults
4. Release *Chrysoperla carnea* predators at 10000/ha to kill all life stages
5. Encourage the activity of predators such as *Encarsia* and predators such as Coccinellids, *Chilocorus nigrita, Chilomenus sexmaculatus*, etc.
6. Spray phosalone 0.07% or triazophos 0.08% or FORS 25g/l or NSKE 5% or acephate 0.11% or TNAU neem oil 0.03% 1ml/l , two to three times based on the incidence.
7. Avoid using synthetic pyrethroids and extending crop growth.
8. Conserve spiraling whitefly parasitoids, *Encarsia haitiensis* and *E. guadeloupae.*

Thrips: *Retithrips syriacus* (Thripidae: Thysanoptera)

- Thrips infest both sides of leaves.
- Infested leaves become discoloured and young plants become stunted.
- In older plants, leaves dry up and fall.

Red spider mites: *Tetranychus urticae* (Tetranychidae: Acarina)

- They cause damage during rainless summer. Mites infest underside of leaves on either side of the mid-rib. Infested regions turn yellowish.
- Attacked plants are stunted. Developmental period varies from 9 to 12 days and adult life from 4 to 10 days. A female lays from 4-26 eggs.
- Mites can be controlled by using acaricides like monocrotophos 0.5% or dicofol 1,5 ml/l or wettable sulphur 2 g/l.

PEST OF PAPAYA AND THEIR MANAGEMENT

1. White fly: *Bemisia tabaci*

Symptoms of damage: Nymphs and adults suck the sap from undersurface of the leaves yellowing of leaves

Identification of pest

- **Egg** - pear shaped, light yellowish
- **Nymph** - Oval, scale-like, greenish white
- Settle down on a succulent part of leaves.
- **Adult** - White, tiny, scale-like adults.

Management

- Field sanitation
- Removal of host plants
- Installation of yellow sticky traps
- Spray application of imidacloprid 200SL at 0.01% or triazophos 40EC at 0.06% during heavy infestation.
- Spray neem oil 3% or NSKE 5%
- Release of predators *viz.*, Coccinellid predator, *Cryptolaemus montrouzieri*
- Release of parasitoids *viz.*, *Encarsia haitierrsis* and *E.guadeloupae*

2. Fruit fly: *Bactrocera (Dacus) dorsalis*

Symptoms of damage

- Maggots puncture into semi-ripe fruits with decayed spots
- Oozing of fluid and brownish rotten patches on fruits.
- Dropping of fruits.

Identification of pest

- Egg - pear shaped, light yellowish
- Nymph - Oval, scale-like, greenish white
- Settle down on a succulent part of leaves.
- Adult - White, tiny, scale-like adults.

Management

- Collect fallen infested fruits and dispose them by dumping in a pit and covering with soil.
- Provide summer ploughing to expose the pupa
- Monitor the activity of flies with methyl eugenol sex lure traps.
- Heavy application of dust and sprays of pyrethrum or BHC
- Spray fenthion 100 EC 2 ml/ lit or malathion 50 EC 2ml/lit.
- Field release of natural enemies *Opius compensates* and *Spalangia philippines*

3. Ash weevils: *Myllocerus spp*

Symptoms of damage

- Grub feed on the roots
- Wilting of young saplings
- notching of leaf margin by adults

Identification of pest

- Grub – small, apodous
- Adult – greenish white with dark lines on elytra

Management

- Collect and destroy the adults
- Spray carbaryl 50 WP at 2g/lit

4. Green peach aphid: *Myzus persicae*

Symptoms of damage

- Nymphs and adults suck the sap from leaves, petioles and fruits
- Leaf curling and falling
- Premature fruit drop

Identification of pest

- Adult: Dark brown to chocolate brown colour

Management

- Remove and destroy damaged plant parts
- Spray dimethoate 0.03% or methyl demeton 0.025%
- Field release of parasitoid *Aphelinus mali* and predators, *Coccinella septumpunctata*

Papaya mealybug, *Paracoccus marginatus* (Hemiptera: Pseudococcidae)

- Commonly called as papaya mealybug
- It is a new record (Jan 2009)
- Exotic in origin
- Invasive on wide variety of commercial crops
- Causing serious economical damage to mulberry
- Affected 1500-2000 acres of mulberry, reduced brushing capacity by 80-90%

Biology

- Small to medium sized, yellow coloured insects with mealy or waxy coating.
- Oval to elongate insects with terminal or waxy filaments. Have well developed legs and antennae
- Eggs are yellow in colour and laid in sac (400 – 500 egg) covered with white wax. Egg period is 7-14 days.
- Nymphs are yellow with 4-5 instars and live for a month.
- First instar nymph is referred to as "Crawler". Upon hatching it moves out and select tender portions and starts feeding.
- Female has four developmental stages (egg – nymph I – nymph II - Adult) and live for about 50-60 days.
- Male has six developmental stages (egg - nymph I – nymph II – pre-pupa - pupa - adult)

- Females are wingless and adult male has a pair of membranous wings; but short lived; die after mating.

Symptoms

- Apical portions are affected initially. Thereafter it spreads all over the plant affecting even woody regions.
- Malformation of affected portion due to toxin injected during feeding.
- Stunted growth of leaf and plant; yellowing of leaves
- Sooty mould on leaves & plants due to honey dew secretions of the pest.
- Movement of ants in the vicinity which help in spread of the mealy bugs.
- It spreads through many ways i.e., planting materials, infested materials, weeds, ants, wind and water.

Host range

- *P. marginatus* is a polyphagous pest and infests almost all plants ranging from Annuals, Fruit crops, Vegetable crops, Weeds, Flower crops and Ornamental crops
- Important hosts are Jatropha, Guava, *Plumeria,* Papaya, Cotton, Pulses, Teak, Tapioca, Sunflower, etc.

Management

- Regular monitoring
- Removal & burning of affected portions to avoid further spread
- Removal of weeds in and around the mulberry garden
- Spot application of pesticides at initial stage of occurrence.
- Immediately after pruning spray 0.2% a.i. DDVP over the pruned shoots and soil around the stem.
- Second spray of 0.1% a.i. Rogor after 10 days of pruning
- Third spray of 0.2% a.i. DDVP or neem formulation @ 3ml / 10 lit mixed in 0.5% soap solution – 10 days after second spray.
- Release coccinellid predators - Cryptolaemus montrozeuri or Scymnus coccivora @ 250 - 300 beetles / acre, a week after second spray.
- **CSRTI, Mysore – Release of solitary nymphal parasitoids like *Anagyrus loecki, Pseudleptomastix mexicana* & *Acerophagus papayae* @ 250-300 adults / acre.**

DISEASES OF CASTOR AND THEIR MANAGEMENT

1. Seedling blight- *Phytophthora parasitica*

Symptoms

- The disease appears circular, dull green patch on both the surface of the cotyledonary leaves.
- It later spreads and causes rotting.
- The infection moves to stem and causes withering and death of seedling.
- In mature plants, the infection initially appears on the young leaves and spreads to petiole and stem causing black discoloration and severe defoliation.

Pathogen

The fungus produces non-septate and hyaline mycelium. Sporangiophores emerge through the stomata on the lower surface singly or in groups. They are unbranched and bear single celled, hyaline, round or oval sporangia at the tip singly.

The sporangia germinate to produce abundant zoospores. The fungus also produces oospores and chlamydospores in adverse seasons.

Favourable Conditions

Continuous rainy weather, low temperature (20-25 C), low lying and ill drained soils.

Mode of Spread and Survival

The fungus remains in the soil as chlamydospores and oospores which act as primary source of infection.

The fungus also survives on other hosts like potato, tomato, brinjal, sesamum etc. The secondary spread takes place through windborne sporangia.

Management

- Remove and destroy infected plant residues.
- Avoid low-lying and ill drained fields for sowing.
- Treat the seeds with Thiram or Captan at 4g/kg.

2. Rust- *Melampsora ricini*

Symptoms

Minute, orange-yellow coloured, raised pustules appear with powdery masses on the lower surface of the leaves and the corresponding areas on the upper surface of the leaves are yellow. Often the pustules are grouped in concentric rings and coalesce together to for drying of leaves.

Pathogen

- The fungus produces only uredosori in castor plants and other stages of the fungus are unknown.
- Uredospores are two kinds, one is thick walled and other is thin walled.
- They are elliptical to round, orange-yellow coloured and finely warty.

Mode of Spread and Survival

- The fungus survives in the self sown castor crops in the off season.
- It can also survive on other species of Ricinus.
- The fungus also attacks Euphorbia obtusifolia, E.geniculata, and E.marginata.
- The infection spreads through airborne uredospores.

Management

Rogue out the self-sown castor crops and other weed hosts. Spray Mancozeb at 1kg/ha or dust Sulphur at 25kg/ha.

3. Leaf blight- *Alternaria ricini*

Symptoms

- All the aerial parts of plants viz., leaves, stem, inflorescences and capsules are liable to be attacked by the fungus.
- Irregular brown spots with concentric rings form initially on the leaves and covered with fungal growth.

- When the spots coaleasce to form big patches, premature defoliation occurs. The stem, inflorescence and capsules are also show dark brown lesions with concentric rings. On the capsules, initially brown sunken spots appear, enlarge rapidly and cover the whole pod. The capsules crack and seeds are also get infected.

Pathogen

- The pathogen produces erect or slightly curved, light grey to brown conidiophores, which are occasionally in groups.
- Conidia are produced in long chains. Conidia are obclavate, light olive in colour with 5-16 cells having transverse and longitudinal septa with a beak at the tip.

Favourable Conditions

- High atmospheric humidity (85-90 per cent) and low temperature (16-20 C)

Mode of Spread and Survival

The fungus also survives on hosts like Jatropha pandurifolia and Bridelia hamiltoniana. The pathogen is externally and internally seed-borne and causes primaryinfection. The secondary infection is through air-borne conidia.

Management

- Treat the seeds with Captan or Thiram at 4g/kg.
- Remove the reservoir hosts periodically.
- Spray Mancozeb at 1kg/ha.

4. Brown leaf spot- *Cercospora ricinella*

Symptoms

- The disease appears as minute brown specks surrounded by a pale green halo. The spots enlarge to greyish white centre portion with deep brown margin.
- The spots may be 2-4 mm in diameter and when several spots coalesce, large brown patches appear but restricted by veins.
- Infected tissues often drop off leaving shot-hole symptoms. In severe infections, the older leaves may be blighted and withered.

Pathogen

The fungal hyphae collect beneath the epidermis and form a hymenial layer. Clusters of conidiophores emerge through stomata or epidermis. They are septate and unbranched with deep brown base and light brown tip. The conidia are elongated, colourless, straight or slightly curved, truncate at the base and narrow at the tip with 2-7 septa.

Mode of Spread and Survival

The fungus remains as dormant mycelium in the plant debris. The fungus mainly spreads through wind borne conidia.

Management

Remove the infected plant debris. Spray Mancozeb at 1kg/ha.

Minor diseases

5. Powdery mildew : *Leveillula taurica*

White cottony growth on the lower surface of leaveswith yellow discolouration on upper surface.

6. Sterm rot: *Macrophomina phaseolina*

Black discolouration appears near base of stem leading to withering and drying.

7. Bacterial leafspot: *Xanthomonas campestris p.v ricinicola*

Water soaked lesions appear, which later become brown and angular with shining beads of bacterial oozing.

DISEASE OF TAPIOCA AND PAPAYA AND THEIR MANAGEMENT

Tapioca

1. Mosaic

Causal Organism:

- Leaves exhibit chlorotic areas - Leaflets are distorted. Lamina size is reduced and malformed.
- Tuber shows splitting.
- The virus is spread by the white fly Bemisa tabaci.

2. Leaf Spot

Causal Organism:

- Spots are light brown to grey - Many spots seen in single leaf
- Conidiaophores and conidia are produced on the spots affected leaves turn yellow and dry.
- Mycelium is hyaline to subhyaline.
- The conidia are needle shaped hyaline and multiseptate.

Papaya

1. Name: Leaf spot

Causal organism:

- The spots are dirty yellow with brown margin. A chlorotic halo is present around the spots. The spots exhibit merger.

- The mycelium is septate, branched and dark coloured.
- Conidia are single celled, hyaline and oval in shape.

2. **Name: Dry root rot**

   Causal Organism:

   - Leaves exhibit dropping and later drying Basal portion of the stem and roots exhibit decay. Bark splits into threads and a large number of sclertial bodies develop in the affected tissue.
   - Mycelium is septate, brown and much branched.
   - The sclerotia are black, spherical to irregular an produced in abundance.

Virus diseases

3. **Name: Mosaic**

   Causal organism:

   - Leaves exhibit mosaic mottling and malformed.
   - Lamina size is reduced.
   - Vector: Aphids

4. **Name: Leaf curl**

   Causal organism:

   - Leaves show curling, crinkling and distortion. Veinclearing and reduction in leaf size is commonly observed.
   - Veins become thickened and turn dark green in colour.
   - Petioles twist into zig-zag manner. Vector: White fly

GRAINAGE TECHNIQUES IN ERI EGG PRODUCTION

Eri Seed Technology

Eri is the only silkworm reared completely indoor among the *Vanya* silks. Further, *Samia ricini* (Donovan) is the single species of saturniidae that has become fully domesticated.

Seed production is the most crucial aspect in sericulture. Cultivated eri silkworm is multivoltine and there is 4-5 overlapping generations in a year.

The farmers of the North Eastern region of India practice eri rearing throughout the year. Eri silkworm is hardy and resistant to diseases in comparison to other silk producing insects.

But the pebrine disease occasionally create problem. Traditional eri rearers give less importance in disease free seed production resulting occasional crop loss or low production. Management of quality seed production is very important by adopting scientific methods in different phases of egg production.

The eri seed production is not uniform in all seasons. Particularly, summer season is not suitable for seed production due to adverse climatic condition. The embryonic development stage in silkworm is very susceptible to environmental conditions i.e. temperature, humidity *etc.* and is greatly influenced by quality seed cocoons.

Further the amount, rate and quality of food consumed by a larva influence the different parameters like growth rate, developmental time, final body weight, survival and reproductive potential as well.

Atmospheric humidity influences directly on the silkworm pupal growth. Fluctuation of temperature and humidity has significant influence on grainage characters, such as, irregular emergence of moth, occurrence of cripple moth and unhealthy moths. The effective improved technique of eri silkworm seed production is most important to fulfill the demand of seed for the eri rearers.

The important areas required to overcome the constraints in enhancing eri seed production, the improved technology of seed production needs to be practiced in farms and farmers levels. The quality seed cocoon, suitable grainage house, optimum temperature & humidity, improved grainage technology, skilled workers are the basic criteria for production of quality eri seed.

We have to follow certain definite steps for qualitatively and quantitatively better eri seed production as follows

Package of practices of eri seed production.

Seed is the back bone of sericulture industry. Rearing of disease free quality seed is utmost important for production of quality cocoons.

Generally rearers are producing their required seed of their own without assorting any scientific approach which causes outbreak of diseases and with poor harvest. Improved technology has been developed by the institute for production of disease free eri silkworm seed.

Eri Grainage house

A Grainage room having a working area of 34 feet length x 18 feet breadth x 12 feet height all round verandah of 5 feet x 6 feet is required to produce 5,000 dfls per crop.

The preferable temperature and humidity is 25 ± 2 °C and relative humidity of 75 ± 5 %. The room should be preferably east facing with provision for proper ventilation. The plinth of the grainage room should be elevated, dry and damp free.

The concrete floor is suitable for grainage room for easy cleaning and washing. Separate space should be kept for seed cocoon storage, oviposition and pupa/ mother moth examination in the grainage house.

The suitable environment of the grainage house is always influenced on the better and optimum moth emergence, pairing, fecundity and hatchability.

A low cost bamboo made grainage house should have concrete floor, mud plastered wall with thatched roof to maintain optimum temperature in all seasons.

- Thatched roof, mud wall and concrete floor. Temperature of 24 - 27°C and Relative humidity of 75 - 80 %.
- Dry and cool with hygienic conditions inside the rearing room.

Appliences and Chemicals for Commercial Grainage operation

To carryout the grainage operation, following equipment and appliances are required for production of 25, 000 dfls per year in 5 grainage operations.

Items	Size/specification	Quantity
Wooden racks	3.5 m x 0.5 m with 3 selves.	6 nos.
Cocoon preservationcages	1.0 m x 0.5 m with bambooor wire mesh.	54 nos.
Kharikas	Thatch grass/ tree twig make	10,000 nos.
Cotton threads	Soft cotton	40 nos.
Eye protecting glasses	Standard	6 pairs
Weighing balance	Electronic 0.1 g to 500 g	1 no.
Musk	Plastic	6 nos.
Bucket	20 litres capacity	5 nos
Bleaching powder	30 % Chlorinated	20 kg
Lime powder	Slaked	40 kg
Max. & Min. thermometer	Digital/manual	1 set
Hygrometer	Digital/manual	1 set
Hand gloves	Rubber	6 pairs
Sieve	Steel with 30 cm diameter	2 nos.

[Table Contd.

Contd. Table]

Items	Size/specification	Quantity
Moth crushing set	Brass made 20 hole set	3 sets
Microscope	Compound	2 nos.
prayer	Foot/hand sprayer	1 no.
Scissor	Steel	4 nos.
Bloating paper	Standard	50 sheets
Formalin	Commercial	2 litres
Slide and cover slip	Standard	20 packets
Examination table	Wooden 2 m length x 1.0 mbreadth x 1 m height	1 no.
Stool	Wooden/steel	4 nos.
Potassium Laboratory grade. 500 g.	hydroxide 4 bottles	/carbonate

Disinfection of Grainage room and appliances

Disinfection is the act of destruction of disease causing pathogens. A disinfectant is an agent that has the capacity to destroy germs or harmful microorganism. To ensure the successful grainage operation, the grainage room and grainage appliances should be disinfected properly.

Disinfection is very important task for controlling the pathogen during grainage operation. In the tropical country like India high temperature and high humid condition make multiplication of pathogen very high and spread the disease quickly in the un-hygienic condition. Following processes are used for disinfection of grainage appliances.

Mechanical

- Sun drying in hot sun 6-8 hours
- Burning with flame gun

Sun drying in hot sun

- Sun drying of grainage appliances in the hot sun for 6-8 hours control the growth fungal mycelia / conidia and others germs or harmful microorganism. This process is simple and recommended for poor farmers.

Burning with flame gun

- Bamboo made moth cage, bamboo or plastic collapsible mountages are generally disinfected with flame gun to control the different types of germs and pathogens.

Chemical

- Different types of chemicals are used as disinfectant in grainage operation like bleaching powder, chlorine dioxide and slaked lime. For the disinfection of grainage appliances, bleaching powder is recommended.

Use of bleaching powder

- Bleaching powder is chlorinated lime and has characteristic pungent smell of chlorine. Its effectiveness is dependent on the level of chlorine in the compound.
- Generally, 30 % chlorinated bleaching powder is useful for disinfection purpose of grainage appliances in eri culture which has strong oxidizing action for germ control. 5 % bleaching powder solution is generally useful for disinfection of grainage appliances or grainage room.

Preparation of 5 % Bleaching powder solution

- For 5% bleaching powder solution, 500 gm of bleaching powder is added with 10 liters of running water. Mixed the bleaching powder thoroughly with a rod and allowed to settle for some time.
- The solution is preferable to be filtered through a layer of thin cloth to avoid larger particle of lime in the solution.
- The sediment of the solution should be discarded and only the supernatant solution are collected and used for disinfection of grainage appliances using spray machine.

Method of using of bleaching powder solution

- Spray the grainage room and appliances with the prepared bleaching powder solution. Wash or Dip the grainage appliances (moth cage, cocoon preservation tray, *kharika* etc.) in 5 % bleaching powder solution.
- After proper disinfection of grainage appliances, the appliances should be dried in sun light before the grainage operation. For the disinfection of floor of the grainage room 5 % bleaching powder solution @ 1 liter solution / 2.5 sq.m. is useful.

- Drenching grainage hall with 5 % bleaching powder solution and finally washing with clean water is essential.
- The whole process should be completed before 3-4 days of grainage operation. Disinfection mask, hand gloves, *etc.* should be used while disinfecting the grainage room or appliances to protect from the health hazard.

Before processing for the seed cocoon ensure that there is no any smell or residual effect of chemical disinfectant (bleaching powder etc.) in the grainage room or appliances.

Slaked lime

It is widely used bed disinfectant and drying agent in sericulture, which has good antiviral activity. It is very effective as dust in improving general hygiene in the grainage operation. Lime stone or $CaCo_3$ is burnt to produce quick lime (CaO) which when hydrated forms slaked lime $Ca\ (OH)_2$ which is used as disinfectant in eri culture.

Method of use

Mix slaked lime with bleaching powder (1:19) and the mixture is used for dusting at the entrance and around the grainage house to maintain dryness with proper hygiene.

Seed cocoon collection and transportation

Seed cocoon should be collected after complete formation of pupae inside the cocoon. Generally in favourable climatic condition, pupation completes within 6-8 days and in winter it takes 8-10 days.

Before harvesting of the seed cocoons, the maturation of the pupae can be checked by cutting some sample cocoon.

Date wise collected ripened worms for mounting should be maintained separately for synchronization of moth emergence and to get maximum coupling.

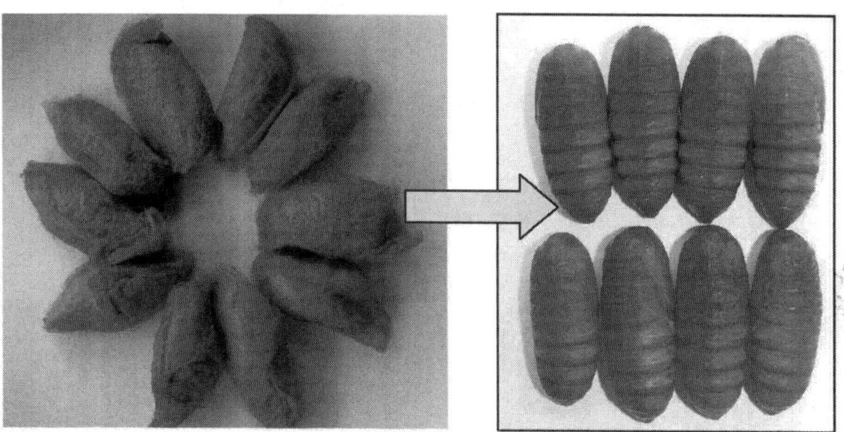

- After harvesting, the cocoons should be transported in bamboo or plastic baskets with sufficient aeration to the grainage room. While transporting the seed cocoons, the following precautions should be observed:
- Seed cocoons should not be transported in sunny hours and exposure to direct sunlight should be avoided.
- Seed cocoon should not be kept near engine of the vehicle.
- Cocoons should not be transported during rains.
- Dropping / falling of cocoons over the hard surfaces and vigorous shaking of seed cocoons should also be avoided.

Selection of seed cocoons

- Before consigning the seed cocoons in the grainage room, seed cocoons are to be examined visually. Stained, deformed or thin layered, pierced, uzi infested and dead cocoons are to be rejected.
- For checking of pebrine disease, microscopic examination of eri pupa is very essential as per recommended method.
- The pupa are crushed in a crushing set (mortar and pestle) adding 5-6 ml of 0.8 % potassium hydroxide (KOH) solution and test under microscope.
- Only healthy, well-built and uniform cocoons are considered as seed cocoons.

Preservation of seed cocoon

- Store the seed cocoons in a single layer moth emergence cages (preferably date wise). Maintain 24 to 26 °C temperature and 70-80 % relative humidity inside the room. Proper aeration is most essential for proper development of eri moth.
- Use of gunny cloth or sand bed on the floor with regular sprinkling of water on the sand bed and spray of lime powder in the grainage hall to control the temperature and humidity is recommended.
- Protect seed cocoons from natural enemies like ants, rats, squirrels etc. Compact grainage room and closed window and door may result emergence of cripple moth, irregular moth emergence or emergence of unhealthy moth.

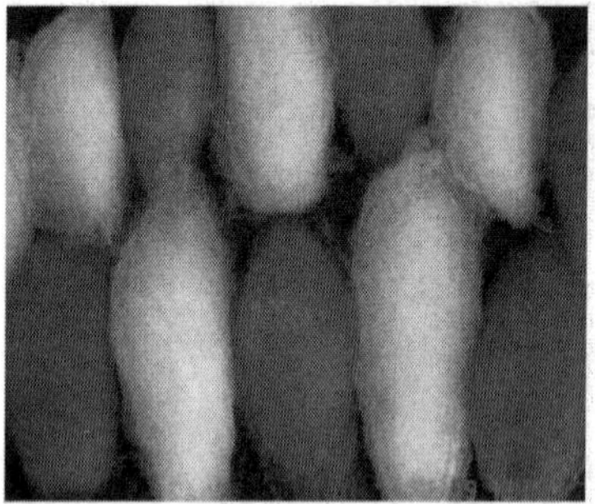

Healthy seeds cocoons

Moth emergence

- Generally, eri moths start emerge early in the morning and at evening and the emergence of moths from cocoons take place after 18/19 days after spinning at 24 – 25 °C.
- However, after the cocoon formation, the pupae undergo metamorphosis for 14 -15 days and develop into moth. The development and metamorphosis process differ in summer and winter seasons.
- The freshly emerged moths are moist, the scaly hair on the soft bodied flexible, wings are small and curled.
- These moths stay motionless for one hour near the cocoon or wall of the moth cage from which the moth emerges.
- The male moths are smaller and active to fly, the gravid females are sluggish and have bigger abdomen with unlaid eggs inside the overiols.

Moth sorting and picking

- The purpose of moth sorting is to eliminate diseased and unhealthy moths, abnormal or underdeveloped wings, scaly hairs dropped and inert, inactive and incapable of copulation to improve the quality of eggs.

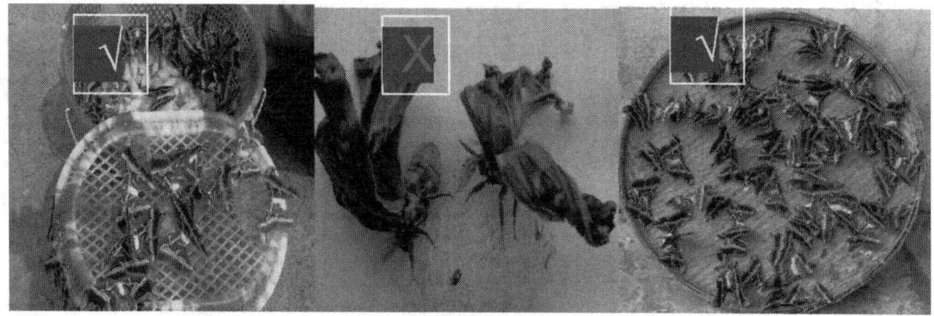

- The healthy moth with bright body, profound wings and large abdomen is considered for pairing and egg production.
- Moth emergence and wing spreading takes some time, after proper spreading the wings, the male and female moths must be gathered separately to prevent injuries each other. By the time deformed and weak moths should be rejected.
- Within a batch of cocoons, a small quantity of exceptionally early or late emerging moths should be eliminated and not included for mating. Healthy male and female moths should be collected in the morning and kept 50:50 ratio of male : female in coupling cages.

Mating of male and female moth

- Eri moths have good coupling aptitude in natural condition, mechanical coupling is not required.
- The optimum temperature and humidity for mating is about 23-24 °C and 75-80 %. Generally, mating takes place after one hour of emergence ensuring 8-10 hours coupling and decoupling is required in the afternoon.

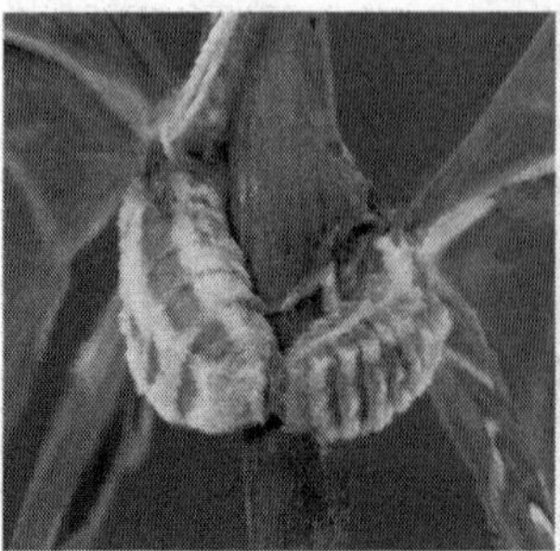

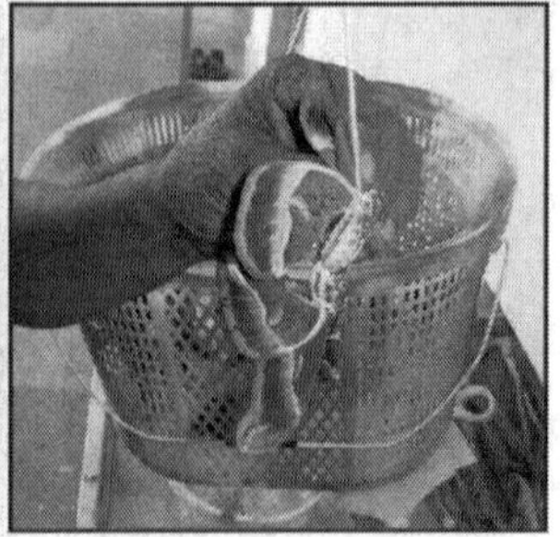

- Male and females moth that emerge in the evening should be kept in pairing cage whole night and collected in the morning for tying.

Picking up and tying of mated female moths

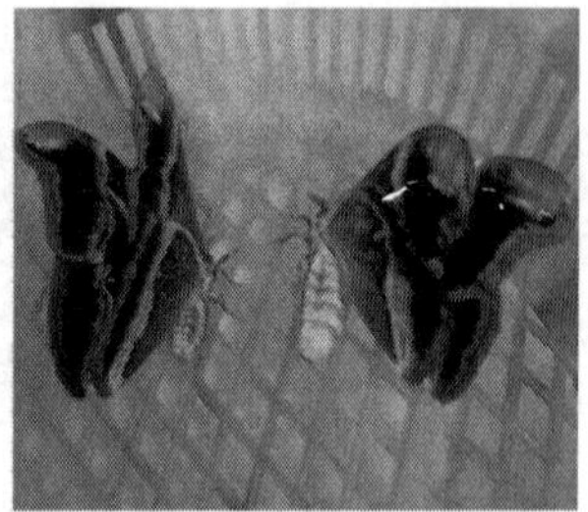

- Within an hour, more than 2/3rd of emerged moths mate naturally. While picking up the mated moths, the base of fore wing of mated female moths is held with thumb and tied it on the *khorika* (small bundles of straw or a split stick 30 cm long with a hook or cut branches of trees having one hook like projection).
- While tying the female moth should be on the upper side and male on the down side. The *kharikas* should be hanged evenly on the string; not to be too closed to make easy separation of mated moths.
- There should be no sound, strenuous vibrations and bright light free in the room. The room should have proper aeration with suitable temperature 24 to 26 °C and 70-80 % relative humidity inside the room and semi dark condition.

Decoupling

Optimum pairing of male and female eri moth is most essential to obtaining the good quality eri seed. Mating for 6-8 hours is sufficient to ensure full fertility. It is observed that eri farmers produce eggs hapahazardly in their traditional practice in the village.

They allow all the moths on the hanging cloth for egg laying, where there is no any time table of coupling and decoupling.

On the cloths some moth remain unpaired and some of the female lay eggs without pairing. The free female moth lay eggs on different places in scattered position, where there is no chance to get individual mother moth examination. In this unscientific process, it is difficult to obtain disease free eggs.

In the scientific methods the mated moths are separated after proper coupling. The separation of mated moths should be done gently holding the wing of female moth and then pulling down the male moth without causing injury to reproductive organs of gravid female.

The separated male moths are kept in the particular moth cage and female moths are left on the *kharika* without removing the thread.

Then the *kharika* is shaken gently to let the female moth pass urine fully. Mother moths not having urinated or not having thoroughly urinated would not only have a late egg deposition, but also would contaminate their eggs when they urinate in course of egg laying.

After proper coupling and decoupling process, the gravid females are ready to lay eggs, hence a shady and cool place is required for oviposition.

In case of shortage of male moths, good male moths are selected, and preserved at low temperature for second time use.

The male eri moth has very good coupling aptitude and hence coupled female should be kept in protective condition, otherwise the active male moth can interfere in time of egg laying.

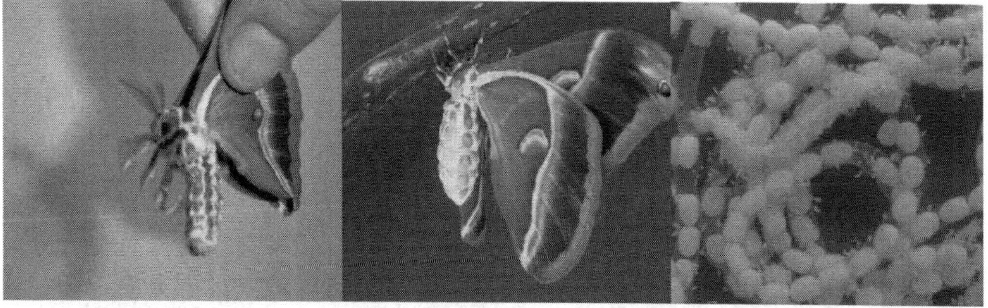

Oviposition

- Eri moths always lay egg comfortably in the vertical position on *kharika* or branch of the tree, where the number of the eggs found more than other oviposition device.
- The suitable temperature and humidity for egg laying is 25-26 °C and 80-90 % in the semi dark condition.

- The decoupled females areallowed to lay eggs on *kharika* in a vertical position and the gravid female start laying eggs in the dark condition from evening hours to late night.
- A nylon oviposition device has been developed but yet to be commercially exploited (Debaraj *et al*., 2008). The female moths lay eggs in clusters.
- The eggs covered with more gummy substances create egg cluster, which indicate that the eggs are healthy. When the egg surface covered with less gummy substances, the eggs are not attached with *kharika* and considered as unhealthy eggs. The unhealthy eggs found in the unsuitable summer season fall down easily from the *Kharika* .
- Eri moths do not properly oviposit in the plain surface or cellule like mulberry silkmoth, If eri moth do not get suitable egg laying device, then number of eggs remain inside the abdomen and the period of egg laying increase.
- They are allowed to lay eggs for a maximum period of 2-3 days. The egg laying capacity of eri moth is different in different seasons. Generally, spring and autumn seasons are suitable for eri seed production than winter and summer. The food plants have a significant role on fecundity of the eri moth.
- The number of eggs laid by one female moth is usually 300 – 350 numbers, but sometimes the number of eggs are found 400 – 450 in the spring and autumn season when worms are fed with primary food plants (Castor, *Ricinus communis* Linn).

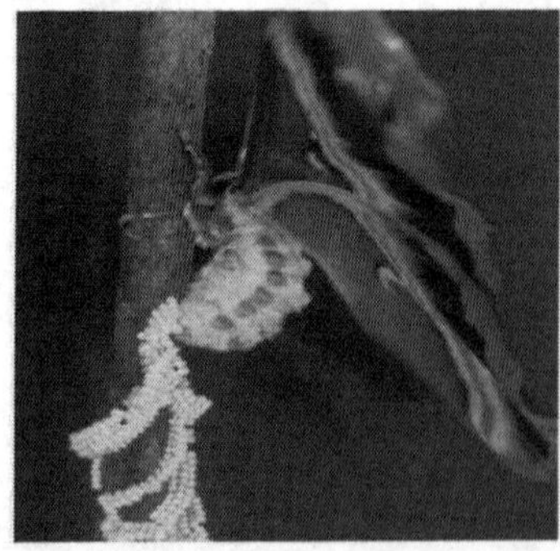

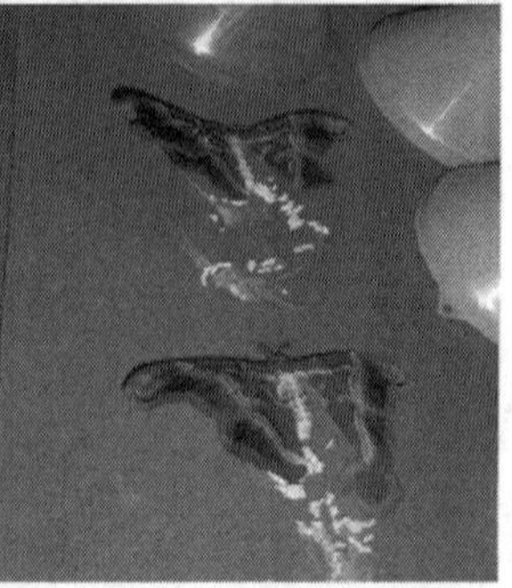

- Pupal weight has significant influences on fecundity, larval weight and other economic characters in the subsequent generation. More pupal weight results in large moth and more egg laying.

Collection of female moths after oviposition

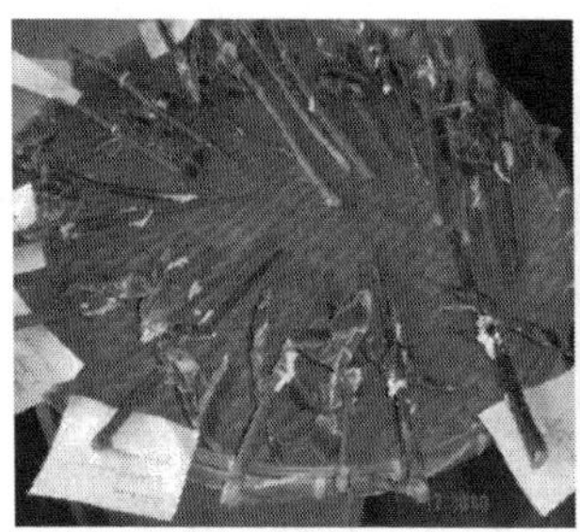

- After 2-3 days of egg laying, the female moths are collected in morning hours and separated date wise for microscopic examination of mother moths.
- The poor layings and unlaid female moths are discarded with the *kharika*. Eri moth lay 90 % eggs within 2-3 days in the suitable environmental condition.
- The life span of the eri moth is 5-8 days depending upon the season. The eggs laid after the 3-4 days are not suitable for rearing.

Mother moth examination

- In order to produce disease free laying (DFL) and to determine the presence of pebrine spores, mother moths are subjected to microscopic examination. Procurement of healthy or pebrine free seed is paramount importance in eri culture.
- It is observed that bacterial, fungal and pebrine diseases are common in eri silkworm. Bacterial or fungal disease can be controlled by surface disinfection and maintaining hygiene condition. But the pebrine is the protozoan disease which infect generation to generation through mother moth. For tackling the problem of pebrine disease, mother moth is examined following individual mother moth examination and mass mother moth examination.

Requirement of equipments/materials

For the improved centrifugal mother moth examination, following equipments / materials are required for detection of pathogen.

1) Mixie with medium size Jars: 1 no.
2) Pastel with grinder: 4 sets
3) Centrifuge (1000 - 5000 rpm) : 1 no
4) Centrifuge tube (50/100 ml): 36 nos.
5) Cyclomixer: 1 no.
6) Plastic beakers: 36 nos.
7) Funnels (10 cm diameter): 36 nos.
8) Scissors: 4 nos
9) Measuring cylinder (500 ml): 2 nos.
10) Muslin cloth: 2 mters
11) Thin glass rod: 5 nos.
12) Micro slide with cover slip: 5 packets
13) Compound microscope: 1 no.
14) Moth examination table and stool: 1 pair
15) Potassium carbonate: 500 g
16) Potassium hydroxide: 500 g
17) Bleaching powder: 5 kg
18) Cotton: 2 roll

Preparation of K_2CO_3 solution

- Add 6 g or 7 g of K_2CO_3 crystal in 1000 ml of water for 0.6 – 0.7 % standard K_2CO_3 solution. The K_2CO_3 solution is very essential to dissolve the fat bodies and tissue of the sample.

Preparation KOH solution

- Add 6 g or 7 g of KOH crystal in 1000 ml of water to get 0.6 – 0.7 % standard KOH solution. The KOH solution is very essential to dissolve the fat bodies and tissue of the sample. In time of the pupal examination the KOH solution is more suitable than K_2CO_3 solution to dissolve the fat bodies and tissue of pupa.

a) Individual mother moth examination

- This is the best mother moth examination process for basic seed production in eri sector. In the method, the abdomen part of the individual mother moth cut by scissors and crushed in a crushing set (mortar and pestle) adding 5-6 ml of 0.6 % K_2Co_3 solution or 0.6 % KOH solution. The smear of the crushed solution can be examined under the compound microscope in (15x40) or (15x45) magnification.
- Now a days, whole moth testing is suggested for detection of pebrine spores. In this method the crushed homogenate are filtered and centrifuged maintaining 3000 to 4000 rpm for 3-4 minutes. The supernatants are discarded and the sediment dispersed in few drop of 0.6 % K_2Co_3 solution and examined preferably in the phase contrast microscope in (15x40) or (15x45) magnification. The moths are tested after 3rd day of oviposition for seed production.

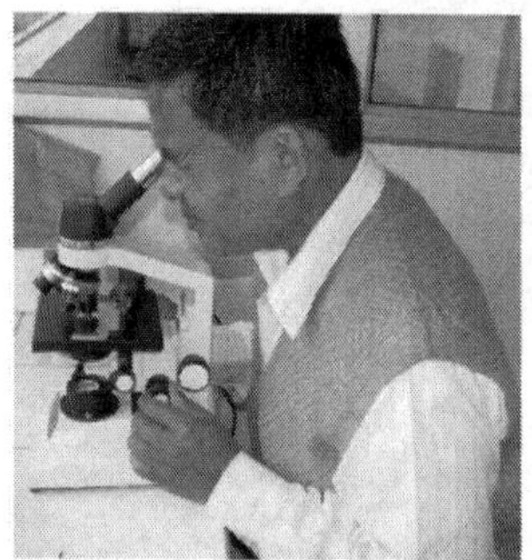
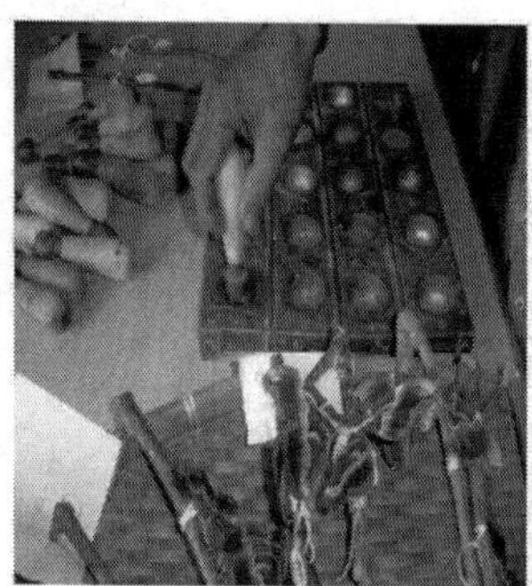
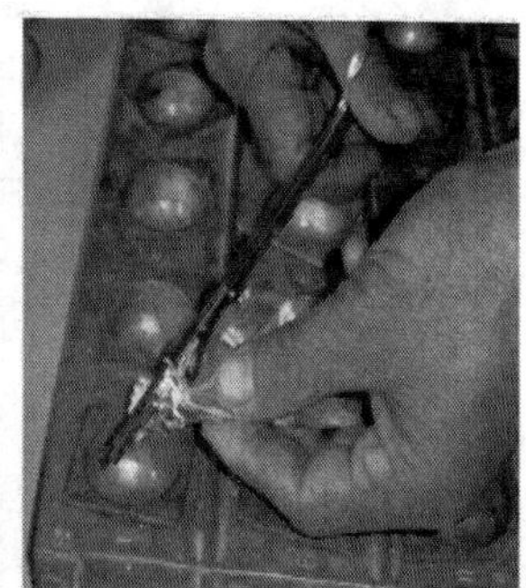

b) Mass mother moth examination

- At commercial level, it is difficult to test individual moth for detection of pebrine spore. Hence a group moth examination system has developed as an efficient method for pebrine detection. Centrifugal method of mother moth examination is effective even when the spore load is low at moth stage.

Method of mass moth examination for pebrine spore detection

1. For 20 nos. of moths 120 ml of 0.6 % K_2CO_3 solution is required.
2. Grind the sample in the grinding machine for 3-4 minute and homogenate for 3-5 minutes.
3. Filter the homogenate through cotton and take the filtrate solution in the centrifuge tube maintaining equal amount in all tubes.

4. Centrifuge the filtrate in 4000 - 5000 rpm for 3-4 minutes for sedimentation.
5. Dilute the sediment with 0.6 % K_2Co_3 solution or in the 0.6 % KOH solution.
6. Mix the dispersed sediment in cyclomixer properly.
7. Take smear from dispersed solution and prepare slide and observe in the compound microscope preferably phase contrast microscope at 600 x magnification.
8. The phase contrast microscope is most preferable for microscopic examination of Pebrine spores.
   - The mother moths infected with pebrine should immediately be rejected along with the *kharika*, cotton thread and eggs laid. It is advisable to burn them to destroy the pathogen. In case of detection of pebrine during egg stage, blue or pin headed stage are more suitable.
   - Disinfection of the mortar and pestle, scissors and other appliances should be done after completion of examination by dipping in 5.0% bleaching powder solution followed by washing in soap and water before reuse. The wings, cut/crushed moths along with debris should be collected in 5.0% bleaching powder and dumped in the soak pit.

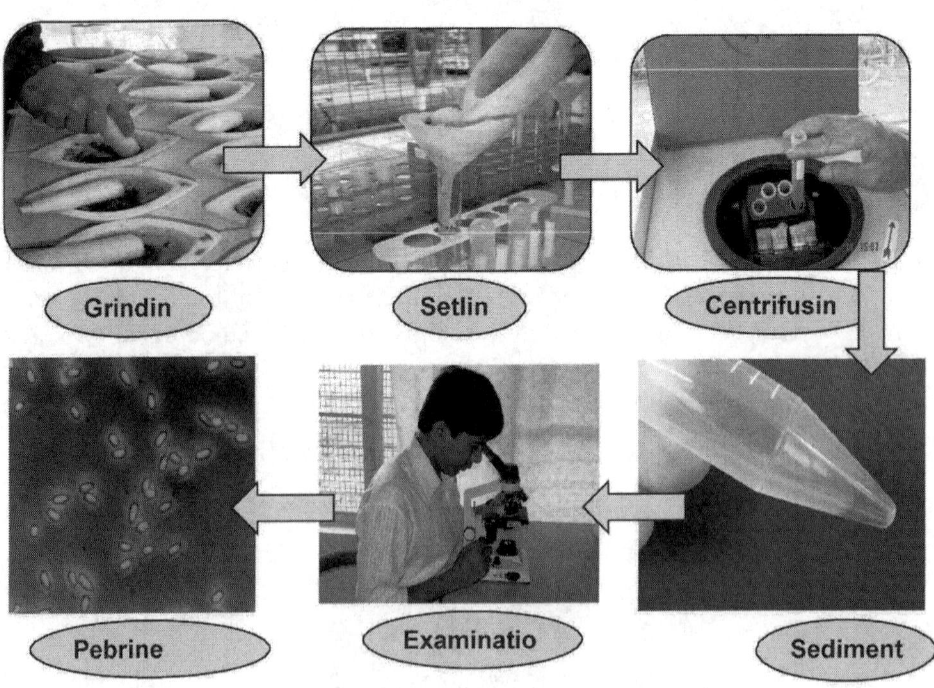

Mother moth examination

- The eggs are removed from the oviposition device after mother moth examination. When mother moth examination is delayed the multiplication and sporulation of pathogen takes place, resulting enhancement of pathogen in the tissue of the moth. Under the delayed mother moth examination, there is more chance to detect the pathogen than the common method of microscopic examination.

Surface sterilization of eggs

- Surface sterilization of eggs should be ensured to avoid contamination. The eggs from the oviposition device after microscopic examination should be collected by dipping in 2 % formalin solution for 30 seconds and then in running water till the traces of formalin is completely removed. Surface sterilization of eggs eliminates fungal infection of eggs.
- Surface sterilization of eggs helps in removal of pathogens adhering to the egg shells and also prevents secondary contamination. The eggs are then dried in the shade spreading in single layer on blotting paper for 5-6 hours in normal room temperature.
- The eggs should not be exposed to direct sunlight/ heat, chemicals etc. to avoid desiccation. After proper drying the disease free layings (dfls) are packed in laying boxes / muslin cloth bags for delivering to the farmers for rearing. Packing should be accompanied with the following details ascribed on a label.
  - ➢ Name of the grainage.
  - ➢ Name of the race / lot number
  - ➢ Either weight in grams or number of eggs
  - ➢ Date of egg laying
  - ➢ Probable date of hatching.

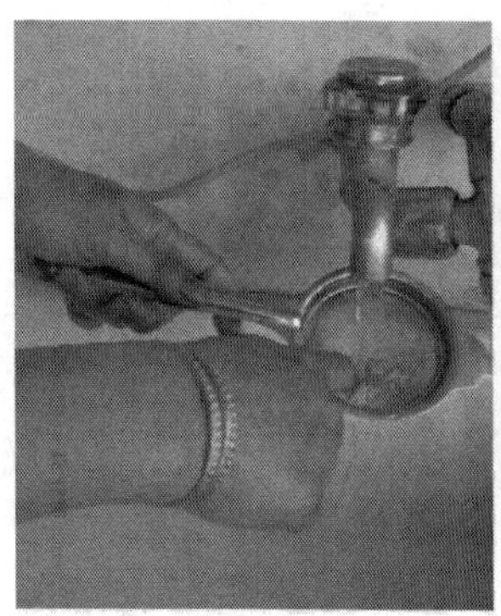

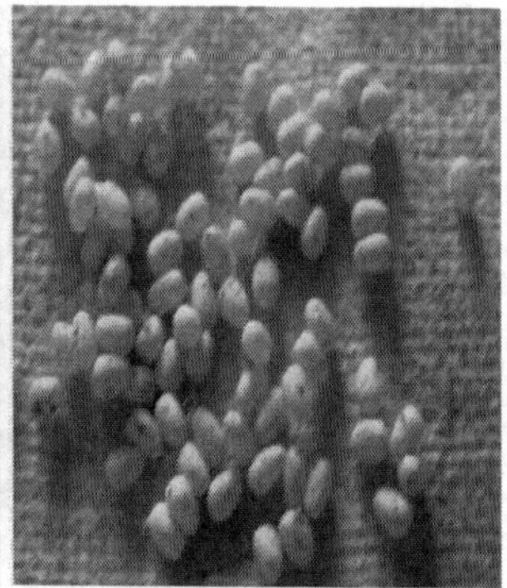

Transportation of dfls

- During transportation, adequate care is required to protect the eggs from exposure to climatic hazards, like high temperature, rainfall *etc*. Avoid stuffing the egg packets inside hand bags or polythene bags and ensure adequate aeration. The following measures should be followed carefully:
  - ➢ Do not expose to direct sunlight, high temperature.
  - ➢ Physical shock on unpaved road should be avoided.
  - ➢ There should not be direct or indirect contact with chemicals, fertilizer, petroleum, insecticide, *etc.*
  - ➢ Do not handle egg with contaminated hands.
  - ➢ Perforated egg boxes should be used.
  - ➢ Do not store eggs in poor aeration condition.
  - ➢ Use moist foam pad at the bottom and sides of the egg carrying busket during hot and dry weather condition.

BIOLOGY OF ERI SILKWORM

- The silk from eri cocoon is not reelable like *Bombyx mori*, the mulberry silkworm, so it is spun into a yarn which is both soft and durable.
- Eri silkworms are continuously brooded, meaning that they do not **go into diapause**, and their life cycle is continuous without regard to seasons.
- The life cycle of eri silkworm is **44 days during summer and 85 days** during winter

Adult

- **Forewings are brownish, blackish or chocolate** coloured with antemedian line and post median line.
- The ante median lime is chocolate coloured with white band on either side.
- The post median line is black with dull grey bands on the side.
- **It also bears the ocellus which is crescent shaped with black margin and bears a hyaline area**
- Fecundity of the moth: **300 – 500 eggs**.

Wing venation

- The anal veins A2 and A3 are fused in forewing.

Wing scales

- Wing scales are generally triangular and bristle like with 1-4 spines at the top.

Life span

- **Male:** 5-6 days normally and it varies based on the temperature conditions.
- **Female:** 10-15 days.
- **The male moth is 2.5 cm long while female is 3 cm.**

Wing span	Fore wing	Hind wing
Male 13 cm	1242 mm2	890 mm2
Female 15 cm	1465 mm2	1037 mm2

Egg

- Freshly laid eggs are slight white in colour.
- As the embryo develops, inside the egg, colour of shell changes from white to yellowish, yellowish to ashy and ashy to blackish just before hatching.
- Eggs are ovoid, candid white.
- **Size of the egg**: 1.5 x 1.0 mm. Weight: 6 mg.

Egg period: 9-10 days.

- On hatching, the larva is greenish yellow in colour.
- Size:5 x 1 mm, weight:1.5 mg.
- Colour changes gradually to pure yellow by end of third day.
- Later the body colour segregates into yellow, cream, green blue or white.
- Mature larva **m**easures 7.0 x 1.5 cm and weights 8 g.
- It is translucent and is covered with a white powdery substance.
- Larva is of 2 types, spotted and unspotted/ plain.

Total larval periods: 17 days - 45 days

Instar	Duration (days)
I	4
II	3
III	3
IV	3
V	6-7

Pupa

- Obtect adectious pupa
- Size: 2.8 x 1.5 cm
- Weight: 2.6 g
- **Pupal period**: 16 days

Biology of *Samiaricini in summer and winter*

S. No	Stage of insect	Life cycle/biology	
		Summer (Minimum days)	**Winter (Maximum days)**
1.	Egg	9	18
2.	Larva	17	45
3.	Spinning stage	3	6
4.	Pupa	13	22
5.	Adult	4	6
	Total	46	97

ERI SILKWORMS – BIODIVERSITY – DISTRIBUTION

Distribution of eri silkworm

- Eri culture was mostly confined to the **Brahmaputra valley** of Assam in the tribal inhabited districts, followed by Meghalaya, Nagaland, Mizoram, Manipur and Arunachal Pradesh.
- Ericulture is introduced on a pilot scale in States like Andhra Pradesh, Tamil Nadu, West Bengal, Bihar, Chhattisgarh, Madhya Pradesh, Orissa etc.
- The Indian Eri Silkworm, *Samiaricini* H. (Lepidoptera: Saturniidae) has several isolated populations, geographically separated (ecoraces) in the states of Assam and Meghalaya.
- Three of them, viz**., Borduar, Mendipathar and Titabar are commercially exploited for the production of eri silk in North Eastern States.**

Table A: Exploited bio-resources of Eri silkworm and their distribution

	Exploited Eri silkworm species	Geographical distribution
a.	***Samia ricini*** **Donovan,** ***Samia cynthia*** **(Drury),** *Samia canningi* (Hutton)	North eastern, India, China, Japan
b.	**Eri silkworm Eco-races** Borduar, Titabar, Dhanubhanga, Sille, Nongpoh, Mendipathar, Titabar, Kokrajhar	North eastern India
c.	**Eri silkworm Strains North eastern India**Yellow Plain, Yellow spotted, Yellow Zebra, Greenish Blue Plain, Geenish blue spotted , Greenish Blue Zebra	North eastern India

Table B: General features of eco-races of Eri silkworm

Ecorace	Location	Larval colour	Cocoon	Silk yield
Borduar	Lower Assam	YP, YZ, GBP, GBZ	White	Higher
Titabar	Upper Assam	YP, YS, GBP, GBS	White	Higher
Dhanubhanga	Lower Assam	YP, GBP	White	Moderate
Sille	Arunachal Pradesh	YP	White	Low
Nongpoh	Meghalaya	YP	Palecream	Moderate
Mendipathar	Meghalaya	GBP	White	Moderate
Kokrajhar	Lower Assam	YP, YZ, GBP, GBZ	Brick red	Moderate

EARLY AGE REARING OF ERI SILKWORM

- The life cycle of eri silkworm is **44 days during summer and 85 days during winter.**
- Rearing is carried out **in doors in a well-ventilated room.**
- **Fecundity of the moth: 300 – 500 eggs.**
- **Life span male: 5-6 days normally and it varies based on the temperature conditions.**
- **Female: 10-15 days.**
- **The freshly laid eggs are slight white in colour.**
- As embryo develops inside the egg, **egg shell changes from whitish to yellowish, yellowish to ashy and ashy to blackish just before hatching**.
- Hatching is seen generally in the morning hours between **7 and 9.0 am**.
- For uniform hatching they should be incubated at **72°F to 80°F.**
- **Egg period**: 9-10 days.
- The eggs on the **stick/ kharikas** or cloth or in a split bamboo are scraped off and tried in a piece of cloth and hung up under the roof until hatching.
- The eggs are to be incubated by spreading then uniformly inside egg boxes in thin layers. The temperature required is **24-26°C and RH is 80-90** per cent.

Brushing

- The cloth containing eggs are opened and placed on a tray.
- Few tender leaves of castor are spread over the worms.
- Larva hatching out during first **two days** is transferred to rearing trays.
- Brushing is done using the tender castor leaves. Avoid crowding during rearing.

Ideal conditions for young age eri silkworm rearing:

- Higher temperature from **26-28°C** and relative humidity ranged **85-90** % are ideal conditions for young age rearing.
- Sufficient tender (glossy) and fresh leaves should be provided to young age larvae.
- The young age worms are very delicate and should be handled with utmost care.

Larval period

Instar	Duration (days)
I	4
II	3

Feeding and its frequency

4-5 feeds per day at regular intervals at 6.0 am, 10 am, 2 pm, 6 pm, 10 pm including the night feeding which is highly essential.

Bed cleaning

- As soon as the larvae grow-up, the unconsumed leaves and litter increase in the rearing bed which ultimately cause changing atmosphere and favoring multiplication of pathogenic organisms.
- Hence, timely bed cleaning is essential to keep the worms healthy.
- Frequent cleaning is better but it involves more labour and ultimately silkworm rearing uneconomical.
- Therefore, it is necessary to prepare stage wise cleaning schedule.
- Only one cleaning is sufficient during first stage worms, otherwise the loss of worms will be more.
- In second stage, two times bed cleaning is required, after first moult and before second moult.
- It should be done after one or two feedings. Cleaning is done using a net.

Space required for 10 layings

Instar	Stage	Space (Sq/ft)
I	BeginningEnd of 1st stage	¾3½
II	BeginningEnd of IInd stage	3½7

Handling of moulting worms and care

- As the worms are entering to moult, stop feeding, lethargic and less movement.
- If 75-80 % of the worms enter into the moult, there is no need to feed the rest of the worms.
- But provide first feeding only when 80 per cent of the worms emerge out of moult.
- Moulting is a very sensitive period during which the worms cast off its old skin and the body is soft and delicate.
- The larvae take **12-36 hours** to complete the moulting process during the different instars and different seasons. It is important to keep the rearing bed dry when the worms are in moult. No bed cleaning should be done during moult.

Application of bed disinfectants

Equipments required for eri silkworm rearing.

Rearing room:

Eri Silkworm should be reared in well ventilated and fly proof rearing room with verandah.

Disinfection of rearing room

- Wash rearing room and appliances with 5% bleaching powder solution.
- Keep the appliances inside the rearing room and seal the room.
- Fumigate 5% Formaldehyde solution under high humid condition or with 2.5 % chlorine di oxide.
- Open the room after 24 hours.
- Disinfect the rearing room at least 3 days before and soon after rearing.
- In case of pebrine incidence the rearing house must be disinfected with 2% formaldehyde.

Race Season

- Rear preferably **Borduar variety.**
- Rearing conducted throughout the year depending upon availability of leaf.

Equipments and uses

1. **Rearing stands**
    - These are frames on which rearing trays containing the silkworms are placed.

- These stands are made up of **wood or bamboo**.
- The bamboo rearing stands are very much popular in **Karnataka and West Bengal.**
- **It measures 2. 25m height, 1.5m. Length and 0.65m width.**
- **It should have ten to twelve tiers spaced at a distance of 0.15m between the two tiers.**
- **Six stands are needed for each** rearing room

2. **Ant Wells**
   - Ant is serious pests of silkworms.
   - They are to be prevented before they crawl on the stand.
   - It is possible by resting the rearing stand legs on ant wells.
   - The simplest ant well is an enameled plate (12 cm wide, 4-5cm deep).
   - **Cement ant wells** (21x21x8cm size with a groove of 4 cm) can also be used.
   - The groove is filled with water to prevent ants on the other hand a piece of cloth dipped in **kerosene** can be placed around the legs of rearing stand.
   - Dusting of BHC also serves the purpose.

3. **Rearing Trays**
   - Rearing trays are portable equipment for keeping silkworms during rearing. There are many kinds of trays differing from one another in materials, shape and size.
   - The selection of the material should not be costly as sericulture is poor man's industry and luxurious equipment are quite out of place. however, the material should not be very flimsy and perishable as it would become costly in long run.
   - In India bamboo round trays are popular as they are light and easy to handle and carry from place to place. It is cheap and can be prepared in every village.
   - It measures 138 cm diameter with a depth of 6.5 cm. **Wooden trays (1.2cm X 0.9m X12 cm)** are used for rearing young stage silkworms. **Two trays** are required till the end of second instar for rearing 100 DFLs. These wooden trays are accommodated in bench type young silkworm stand (wood). The trays are piled one over the other.

4. **Paraffin Paper**
   - This is a thick craft paper coated with paraffin wax with a melting point of **550C.**
   - It is used in modern methods of chawki worm rearing.
   - It is to cover the silkworm's bed to prevent leaf **withering and to maintain required humidity.**
5. **Foam Rubber Strips**
   - For modern method of rearing long foam rubber strips (2.5 X 2.5 cm size) are essential.
   - These are dipped in water and kept around the chawki worm bed to maintain optimum humidity. As a substitute, newspapers folded into convenient strips are also used.
6. **Chopsticks:** This bamboo stick is **17.5 to 22 cm long,** thin in girth and tapering at one end. These are used for **pocking the young larvas**. This ensures hygienic handling of delicate young worms
7. **Feather:** A white quill (long) feather is most essential in the rearing room. It is used for brushing newly hatched worms and also for changing beds in the early stages. It is the most convenient and safe for brushing
8. **Chopping Board, knives and mats**
   - The mulberry leaves are chopped according to the age of the worm before feeding.
   - This equipment is essential for chopping the leaf in a proper way.
   - The board is made up of soft wood (92X92 cm with 7.6 cm thickness).
   - Each rearing house needs two knives (small one for chawki rearing, big one for late age rearing).
   - The knife is 0.3 to 0.5 m long blade with 4 to 8 cm breadth.
   - Leaf chopping is carried on the mats to avoid dust, dirt and helps to collect the chopped leaves.
9. **Leaf Chamber**
   - The rearer cannot go to field to get the leaf for every feeding. And it is also not advisable to cut the leaves when it is too hot or raining. The required leaf is therefore plucked by hired labour in the early hours or late in the afternoon.
   - Thus it is necessary to preserve the nutrients of mulberry leaf till feeding. It can be achieved by preserving the leaf chamber/eartherned pots/ refrigerators.

- A simple leaf chamber is made up to wooden strips (152 X 76 X 76 cm size;strip size is 7.5 cm fitted with 7.5 cm spacing).
- The chamber is covered on all sides with gunny cloth which is made wet.
- During summer season and dry days water is sprayed periodically on the gunny cloth.
- The leaf can also be preserved in earthened pots.
- The rearer should remember that heaping of leaves in a corner is worst.
- Because the leaves rapidly wither owing to evaporation especially in dry season.
- On the other hand temperature inside the heap raises favouring the fermentation in the leaves. Besides this there is every danger of the leaves getting infected with diseased germs, dust especially during cleaning, bed changing etc. thus it is necessary to preserve the leaf in leaf chamber

10. Cleaning Nets

- During the course of feeding certain amount of mulberry leaf is left behind in the tray. This dried leaf together with litter of the worms accumulates in the rearing beds, owing is increase of bed size.
- Further is also adds to increase temperature and harmful gases in the bed. Thus the waste leaf as well as litter has to be periodically removed using nets, basing on the age of the worms.
- The nets are woven using cotton or nylon thread. The mesh should be square and of the same length of the silkworm.
- The rearer should possess small (2mm2), medium (10mm2) and large (20mm2) nets (for rearing chawki and late age worms.)

11. Mountages

It is used to enable the ripe worm to spin cocoon. The most, common form of mountage in India is "Chandrika".

It is a rectangular bamboo mat on which a spiral bamboo tape is tied. The chandrika measures 1.8 m X 1.2 m. The tape is about 4-5 cm. broad and space between the spirals is about 4-5 cm.

Besides chandrika there are other types of mountages which are in practice i.e., zig-zag mountage, centipede mountage, rotary mountage etc. Zig-zag mountage frame is made of plastic wire or sedge straw/mulberry branches/ bamboo, usually consists of about 40 corners of about 6 cm each, and each set can mount **250-300 worms.**

In **India straw mountages** in Assam, **mulberry branch mountages in Jammu & Kashmir are in practice. Recently P.V.C. mountages are**

also used. It has the same advantages like chandrike besides reducing the labour expenses.

Rotary mountage has pieces of cardboard to form **13 rows, consisting of 12 sections and each and amounting to 156 sections.** Ten pieces are put into frame as a set when this frame is hung up with wire holding at by ends, the frame can be turned around two axes. This frame produces fewer spoiled cocoons and raises the reelability of cocoons. Good cocoon percentage is more than 80. It is the best type and suitable for large scale silkworm rearing because it does not only lead to increased cocoon quality but also saves labour in mounting and harvesting.

Centipede mountage is made using plastic or straw material. It provides for **self mounting similar to rotary type.** But care should be taken to avoid a high density of worms on the mountage because it may cause a higher incidence of bad quality cocoons i.e. double and deformed cocoons. It is easy to transport and low cost.

12. **Foot Cleaning Tray**

- Silkworms are very sensitive to dust and microorganisms which may cause damage. Thus they are prevented prior to their entry into rearing room along with foot of rearer.
- This can be achieved by keeping foot cleaning tray at the entrance of the room. It measures 1.1m X 0.5m X 0.1 m with tin bottom.
- A pad of gunny cloth soaked in 2% formalin in kept in the tray. Thus, the persons entering into the rearing room disinfect their foot.

13. **Basin & Stand:** While entering into the rearing room it is advisable to disinfect hands before handling the worms. Thus basin, stand (tripod) is required to keep disinfecting solution.

14. **Feeding Stand:** At the time of feeding the worms, wooden feeding stands are required for keeping the trays, (90X60 cm with 5X2 cm tape). It can be folded.

15. **Other Appliances**

- For reading temperature and humidity in the rearing room dry wet thermometer, hygrometer, humidity chart are essential.
- During winter especially in Kashmir, West Bengal it is necessary to heat up the rearing room to increase the temperature. Thus, an electric, charcoal stove is essential. On the other hand it is necessary for fumigation process also.

16. Sprayer

- Sprayer is required for disinfecting the rearing room, equipment and also spraying water on the leaf chamber.
- Leaf baskets to carry leaves from the mulberry field to rearing house, litter basket to collect waste leafage, litter etc; disinfection mask for workers remaining in the rearing room for longer period for effective disinfection; black box/paper/cloth for black boxing of eggs; humidifier for increasing humidity are also required for rearing room.
- For measuring chemicals/formalin and to weigh the leaf for assessment of growth a rough balance, measuring cylinder, buckets are required. Other items like gunny cloth, uzi proof net are also required.

17. Chemicals and uses: For successful rearing of silkworm maintenance of hygienic conditions are necessary. To prevent and control the microorganism's disinfectants like formaldehyde**,** paraformaldehyde, bleaching powder, sodium hypochloride, slaked lime powder is necessary. Silkworms are not so exception for diseases. Thus as a preparatory step Reshamket Oushadh (RKO), Vijetha uzicide, and fungicides like Dithane M-45 or Captan are kept along with other chemicals.

LATE AGE REARING OF ERI SILKWORM

Late age silkworm rearing

- A late age worm consist of third, fourth and fifth instars which need preferably lower temperature and humidity during rearing conditions.
- These stages consume more quantum of food than the young age worms because the worm has not only to develop silk glands and to increase growth rate, but also has to store up the reserve food materials for the future stages like pupa and imago. Therefore, these stages should be provided as much as quality food they require.
- The late age worms consume 80-85 % leaves supplied during the entire larval period. They attain significant growth during this stage. If the chawki rearing is conducted perfectly resulting healthy and robust worms with less mortality, the late age worm rearing will be easy with more chances of a successful crop.
- But proper care is essential to obtain the full potential of larval growth, maximum yield and best cocoon quality providing with sufficient food. The environmental and nutritional conditions required in late rearing are different from that of young age.
- The ideal conditions of temperature ranged **24-26**°C and relative humidity ranged **70-80 %** should be maintained during the rearing of late age worms. These stages should be provided with **semi-matured leaves**.
- The feeding of dried, yellow and diseased leaves deteriorates the health of the worms and even death due to diseases. Since 80-85 % of leaves are consumed by these stages, ensure continuous supply of quality leaves after preservation in the leaf chamber.

There are three methods of eri silkworm rearing, i.e.,

i. Bunch- rearing,

ii. Tray – rearing and

iii) plat from rearing.

- The farmers of the northeastern region generally employ both the methods either singly or in combination together.
- **The new recommendation is that the first three instars should be reared on trays, and the fourth and fifth instars** on the bunch.

1. Bunch rearing

- In bunch rearing method, 10-12 leaves of castor or branches of kesseru are tied together to make a bundle and hung vertically on a horizontal bamboo or wire or string for support. Then the worms are allowed to feed on the tied leaves.
- The foliage is changed by keeping fresh bunch near the exhausted one and the worms crawl over the new one. Just below the hanging bunches bamboo mat or tray is kept on the floor so that the worms which fall down are not contaminated with dust on the floor and can be picked-up and put on the bunches.

2. Tray rearing

- In tray rearing method, the worms are reared providing the leaves on the tray. Trays are made up of either bamboo or wood in different shapes and sizes. The shapes are round (bamboo made) square and rectangular (wooden).
- The young age (I-III instars) silkworm rearing is conducted either in the wooden trays of 50 cm x 60 cm x 5 cm size or in bamboo tray (of 70 cm dia.). However, bamboo trays of size 1.0 m diameter is more convenient to rear 10-15 dfls until 2nd / 3rd instar; while 600-700 worms can be reared up to 4 th instar and 300 worms in the final instar which also provides sufficient space.

3. Plat form rearing

- In Platform rearing method, platform is prepared using wood or bamboo. Shelves are arranged in two tiers with an interval of 30 inches or in three tiers with an interval of 27 inches in between the tiers.
- The rearing seats of the shelves are prepared by using nylon mesh or bamboo mats. It is generally recommended for **mass scale rearing** (not less than 100 dfls) of late age eri silkworms. The ideal size of the platform is 5’ width and 7’ length. Length can be extended with the availability of the space and rearing capacity.

Maintenance of larvae

During the entire rearing period, the stage wise general maintenance of the leave is required considering the following aspects,

a) Feeding and its frequency,

b) Bed cleaning,

c) Spacing of worms,

d) Handling of moulting worms and care,

e) Collection and destruction of weak, diseased and undernourished larvae.

Feeding and its frequency

- In late age worms due to increase feeding capacity, 5 feeding per day are essential.
- Five feeds per day at regular intervals at 6.0 am, 10 am, 2 pm, 6 pm, 10 pm including the night feeding which are highly essential.
- In the night time more than sufficient leaves should be provided to fulfill the required consumption throughout the night.

Bed cleaning

- As soon as the larvae grow-up, the unconsumed leaves and litter increase in the rearing bed which ultimately cause changing atmosphere and favouring multiplication of pathogenic organisms.
- Hence, timely bed cleaning is essential to keep the worms healthy. Frequent cleaning is better but it involves more labour and ultimately silkworm rearing uneconomical
- Three bed cleanings should be resorted to the third and fourth stages, first after second moult, second in the middle and the third before third moult, similarly in between third and fourth stage.
- But in fifth stage, daily bed cleaning is essential preferably in the morning after one or two feedings. The method of bed cleaning practiced in ericulture is simple and easy. Prior to bed cleaning, a feeding should be the new foliage. Then the worms along with the new foliage should be transferred to a new rearing tray carefully

Spacing of worms

- The maintenance of optimum number of worms per unit area according to the size or stage of the worms during rearing is called spacing of worms.
- Proper spacing and good aeration keeps the worms healthier.

- Overcrowding of the worms in the tray leads to competition for food and space and ultimately undernourishment and unhealthy growth of the larvae often resulting in crop loss.
- Optimum spacing is, therefore, to be accomplished through experience

Space required for 10 layings

Instar	Stage	Space (Sq/ft)
III	BeginningEnd of IIIrd stage	716
IV	BeginningEnd of 4th stage	2640
V	BeginningEnd of 5th stage	4080

Handling of moulting worms and care

- A good rearing is judged by the uniformity of the larvae entering into moulting and emerging from moulting.
- The brushing and feeding of the worms play key role for uniform moulting, if handled properly. As the worms are entering to moult, stop feeding, lethargic and less movement.
- If 75-80 % of the worms enter into the moult, there is no need to feed the rest of the worms.
- But provide first feeding only when 80 p.c. Moulting is a very sensitive period during which the worms cast off its old skin and the body is soft and delicate
- The larvae take **12-36 hours** to complete the moulting process. It is important to keep the rearing bed dry when the worms are in moult. No bed cleaning should be done during moult. After completion of moult, the rearing bed has to be disinfected with bed disinfectants like, Vijetha, Ankush, Suraksha at the rate of **4 kgs/100 dfls.** Feeding is given 30 minutes after bed disinfection.

Collection and destruction of weak, diseased and undernourished larvae

- During rearing, the weak, injured, diseased and irregular worms should be collected immediately and put in **2 per cent formalin solution.**
- Such worms should be buried or burnt carefully to prevent spread of diseases. If rearing continues mixing with these worms, the losses due to contamination and disease spread will be more; even crop failure may take place.

Matured worm collection and mounting

- After completion of larval life span, the matured 5th stage larvae discarded its complete excreta consisting of liquid and semi-solid substances. Now the worms are ready for spinning cocoons. Before spinning, the worms stop feeding and become restlessly moving here and there to search a suitable place for cocooning. Worms become smaller, flabby.

- The matured worms produce a hollow sound when it is rubbed gently between fingers. This is the time for picking the ripe worms and putting them on mountages. Before mounting process, the required number of mountages should be kept ready well in time. The worms should be collected carefully for mounting. While cocooning, it is observed that the worms require at least two supporting sides. The size of the cocoon depends upon the size of the space available for cocooning.
- The commonly used mountages are **chandriki,** basket filled with dry leaves, jali (a bundle of dry leaves like mango, jack fruit, some ornamental plants, etc.) and gunny bag filled with dry leaves.
- The leaves should not be completely dried; semi-dried leaves are suitable for easy spinning. After keeping the optimum number of worms in the respective mountages, it is covered by newspaper or cloth to make support and calm and semi-dark, a suitable condition for cocooning is complete. The ideal condition for spinning is around 24-25°C temperature and 60-70 % relative humidity.
- Due to unavailability of suitable place and disturbance, the larvae spin a defective cocoon and even fail to spin cocoons. In some cases, the physiological condition inside the body is abnormal and the larvae often die without pupating.
- Mounting of immatured and over matured larvae results in poor cocoon quality. The quality of cocoon is also depending upon the type of mountages, density of worms in mounting and different mounting methods / models.
- During spinning, temperature, relative humidity and aeration influence cocoon quality. The ideal condition for spinning is around 24-25°C temperature and 60-70 % relative humidity.
- While mounting, the optimum number of worms should be maintained per mountage, *i.e.,* **300 worms per chandraki of 1.0 m diameter size.**

Harvesting and assessment of cocoons:

- After completion of spinning, the larval skin is cast off and pupation takes place. The pupa has a thin cuticular skin which is soft and may get ruptured easily, if disturbed. The last and important step is harvesting of cocoons from the mountage in time.
- Cocoons should be harvested after **5-6 days of spinning in summer and 8-9 days in winter.** The harvesting process is the best time of sorting of cocoons according to the quality. The cocoons should be sorted out into good, double, melted, stained, dead or inferior, cut or pierced cocoons.

- Good commercial cocoons should be shifted and dried perfectly after the harvest. Cocoons should be preserved carefully to protect from fungal infestation and attack from pest and predators. Cocoons should be assessed on the basis of cocoon weight, shell weight and silk ratio.

Transportation of cocoons

- Transportation should be in the cooler hours of the day.
- Cocoon should be packed in bamboo basket with shelves to ensure free aeration, and to avoid damage due to overlapping.

Rearing schedule for Eri Silk worm:

1. Spring Season

Name of crop	Date of hatching / Bushing	Date of maturing Spring season	Date of emergence
1st batch	March-16 to 18	April-8 to 10	April-21 to 23
2nd batch	May-4 to 6	May -22 to 24	June-5 to 7
3rd batch	June-16 to 18	July-4 to 6	July-17 to 19
4th batch	July-28 to 30	Aug-16 to 20	Aug-31 to Sep-5

2. Autumn Season

Name of crop	Date of hatching / Bushing	Date of maturing Spring season	Date of emergence
5th batch	Sept-14 to 17	Oct-16 to 19	Nov-3 to 6
6th batch	Nov-18 to 20	Dec-23 to 25	Jan-15 to 17
Basic seed rearing	Jan-27 to 28	Feb-19 to 21	March-6 to 8

ERI SILKWORM REARING MANAGEMENT

- The domesticated variety of eri silkworm, *Samia ricini* (Donovan) is multivoltine in nature.
- The structure of the genitalia, wing pattern and chromosome number demonstrates that *Samia ricini* (Donovan) is derived from its wild form, *Samia canningi* (Hutton).
- Several eco-races like Borduar, Titabar, Khanapara, Nongpoh, Mendipathar, Dhanubhanga, Sille, Kokrajhar, Diphu, Genung etc. (Chakraviorty *et al.*, 2008). of eri silkworm are available in North Eastern region of India.
- Depending upon larval colours and markings, six pure line strains were isolated from Borduar and Titabar eco races like **Yellow Plain, Yellow Spotted, Yellow Zebra, Greenish Blue Plain, Greenish Blue Spotted and Greenish Blue Zebra**. Eri silk is unique among other silks for its typical quality of white soft yarn possessing thermal properties.
- In sericulture industry, rearing is the most important and critical phase. Silkworm rearing depends upon the prevailing climatic conditions of the place of rearing, availability of essential facilities/ materials like food plants, rearing house, appliances, equipments, etc.
- The eri silkworm is reared indoors like mulberry and can be reared five to six times all through the year subject to availability of food plants. Unlike other silkworms, eri silkworm rearing is simple and does not require high skill. Eri silkworms are hardy and less susceptible to diseases.

Different stages of eri silkworm

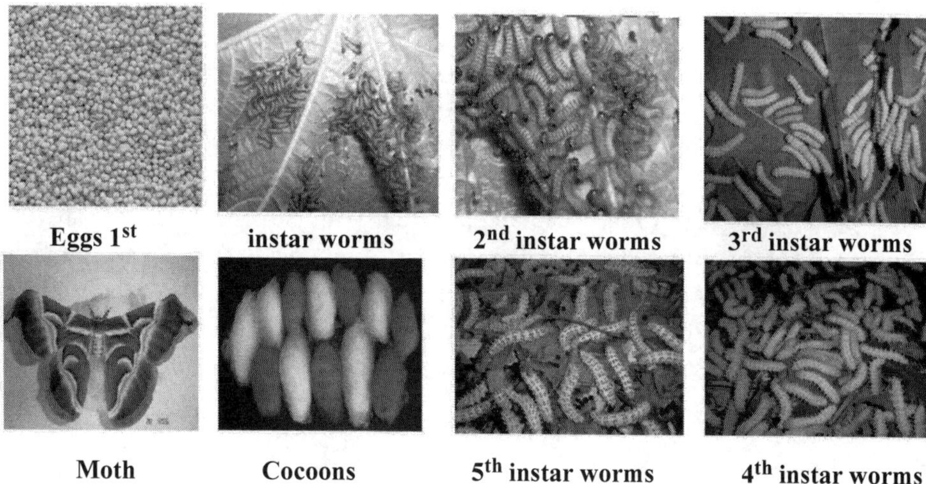

Eggs 1st instar worms 2nd instar worms 3rd instar worms

Moth Cocoons 5th instar worms 4th instar worms

- The crops are assured as compared to mulberry, muga and tasar. However, the present trend of eri cocoon production at farmers' level is below expectation and far behind the production potential of eri silkworms.
- It is mainly due to lack of technical knowledge, non-availability of the essential infrastructures and mismanagement during rearing.

Therefore, well planned, managed and fulfillment of all the pre-requisites for rearing are essential to boost-up the cocoon production qualitatively as well as quantitatively. The main objective of silkworm rearing is to get high profit, which can be achieved by producing good quality cocoons in large numbers. In eri culture, the cocoons are meant for silk production and the pupae for seed as well as a protein rich diet for the people of North East India. Apart from the availability of good quality silkworm seeds, proper disinfection and maintenance of hygienic conditions of rearing room, appliances, maintenance of optimum temperature and humidity, availability of adequate quantity of quality leaves and man-power during rearing are the major contributing factors towards the success of silkworm rearing. The arrangement and procurement of all the pre- requisites of rearing is the first and fore-most task prior to rearing. These include rearing house, food plants, rearing appliances, selection of race and season, number of crops per year, etc. The following are the different aspects of the eri silkworm rearing technology and its management (Sarmah *et al.*, 2012).

Package of practices of improved eri silkworm rearing and its management.

Availability of food plant

A rearer has to decide the size of the rearing to be conducted after estimating the availability of eri food plants. The most common eri food plants are castor, kesseru, tapioca and payam, which are abundantly found in North Eastern region of India. Another perennial food plant Borpat, *Ailanthus grandis* is a promising eri food plant. Out of these, castor is the most preferred food plant of eri silkworm and it can be utilized throughout the year. The other food plants can be utilized in suitable seasons or interchanging among others. Maintenance of own plantation plot is essential for an effective rearing.

Rearing House

Eri silkworms are reared indoors. The plinth area of 10 m x 5 m size rearing house having tin or thatch roofing with 1.5 m verandah all around is ideal for accommodating 100 dfls for commercial and 50 dfls for cellular stock maintenance of eri silkworm rearing per crop. The rearing houses are to be built with adequate number of windows for maintenance of a good environment and ventilators which are also to be fitted with nylon nets to stop entry of uzi flies. Efforts should be made to maintain temperature, relative humidity, light and other hygienic conditions inside the rearing house in different seasons. Plantation of evergreen trees around the rearing house should be encouraged for maintaining a better environment. The rearing house should have enough space for leaf preservation, young age and late age silkworm rearing and mounting. It should have the facility for disinfection and cleaning conveniently.

Requirement of rearing appliances

All the required appliances and materials for rearing should be made available prior to rearing. These materials should be kept clean and disinfected properly. The following major items are required for rearing 100 dfls of eri silkworm.

Name of the items	Specifications	Quantity
Hand sprayer	Plastic	1
Rearing stand (wooden)	2 m x 1.8 m x 0.50 mwith 5 shelves	10

[Table Contd.

Contd. Table]

Name of the items	Specifications	Quantity
Feeding stand (Wooden)	1.0 m height	10
Rearing tray (Bamboo)	1.0 m diameter	100
Chandrika (Bamboo)	1.0 m x 1.0 m	100
Ant wells (Aluminum)	15 cm diameter	40
Disinfection mask	Standard	3
Leaf preservation chamber (Wooden)	1.5 m x 1.0 m x 0.75 m	2
Egg	hatching	box (Plastic)
0.60 m x 1.0 m	100	
Bucket	20 lit capacity	2
Wash basin	Standard	2
Wash Basin stand	1.0 m height	2
Plastic mug	1.0 lit	2
Hygrometer	Digital or wet & dry bulb	2
Max-min thermometer	Standard	1
Foam pad	40 cm x 60 cm	2
Measuring cylinder	1lt	1
Basket (Bamboo)-	—	1
Old news paper	—	5 kg
Bleaching powder	—	5 kg
Bird feathers	—	10
Slaked lime	—	5 kg

Disinfection and prophylactic measures

Complete and thorough disinfection of rearing house and appliances is vital for successful rearing. In fact, disinfection, before and after each rearing is considered the key for a successful cocoon crop.

To protect from pathogens, special attention is needed for disinfection of every nook and corner of the rearing house and appliances with proper chemicals in correct concentration. Disinfection should be carried out on bright sunny days. Some commonly used disinfectants in sericulture are formalin, bleaching powder, lime, sodium hypochlorite and chlorine dioxide.

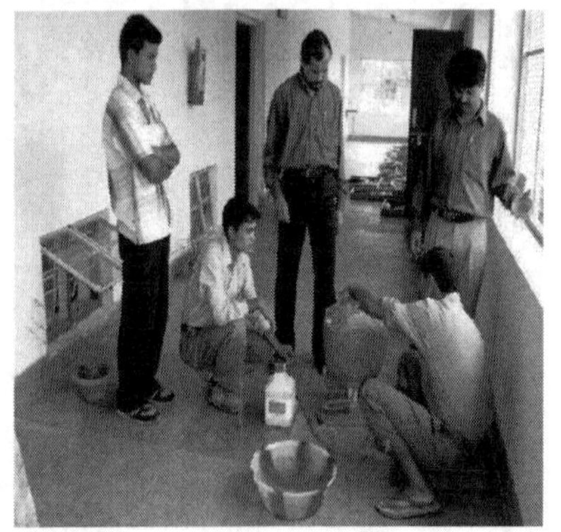

The disinfection with 5 % bleaching powder solution is effective. Sprinkling of 2 % bleaching powder-lime mixture in the surroundings of the rearing house from time to time is good. The room temperature should be maintained at around 25 °C during disinfection. All the crevices and holes of the room should be closed to prevent entry of pests, predators, pathogens *etc*. Windows and ventilators should be kept open for proper aeration and free circulation of air.

Selection of races and season

So far, 25 ecoraces of eri silkworm are characterized collected from different parts of North Eastern region of India, *viz*., Borduar, Titabar, Khanapara, Mendipathar, Dhanubhanga Kokrajhar, Nongpoh, Diphu, Borpeta, Imphal, Inao, Mukokchung *etc*. (Chakravorty *et al*., 2008 and Sarkar *et al*., 2012) In addition, six strains are isolated and maintained in the germplasm bank of Regional Eri Research Station, Mendipathar in Meghalaya and Central Muga Eri Research and Training Institute, Lahdoigarh. These include yellow – plain, yellow – spotted, yellow – zebra, greenish – blue plain, greenish – blue spotted and greenish – blue zebra (Debaraj, *et al*., 2001). Borduar and Titabar eco-races are better yielder among the ecoraces of North Eastern region of India (Sarkar *et al*., 2012).

Different strains of eri silkworm

The best season for eri silkworm rearing is June-October during which the rearing performance is found better (Sarmah et al., 2012). . The new eri breed C2 has been developed by hybridization of two potential parents SRI-018 (Genung) and SRI-001 (Borduar) through exerting directional selection at Regional Eri Research Station, Mendipathar, under Central Muga Eri Research and Training Institute, Lahdoigarh, Assam. (Singha, 2010). C2 breed shows best performance on feeding of non-bloomy red variety of castor (NBR-1). However, it can be reared on Kesseru (*Heterpanax fragrans*), Borpat (*Ailanthus grandis*) and Borkesseru (*A. excelsa*). The prevailing climatic conditions of the North Eastern region of India are congenial for eri culture and eri silkworms can be reared in 4 – 5 overlapping crops in a year.

Castor (NBR-1) **Borpat**

Borkesseru **Kesseru**

Maintenance of environmental conditions during rearing

The maintenance of ideal environmental conditions during different stages of rearing has a significant influence on the larval growth and ultimately a good crop.

A fine weather environment refers to individual factors such as temperature, humidity, air and light. The influence of all the factors varies in different stages of larvae. It is maximum in the first instar and minimum during the fifth instar. Care should be taken not to expose the worms to extreme climatic conditions for a long period. High humidity keeps leaf fresh and silkworm feeds well but helps in outbreaks of silkworm diseases. The temperature requirement during the early instars is high and low in the late instars.

Egg incubation

Egg incubation and hatching

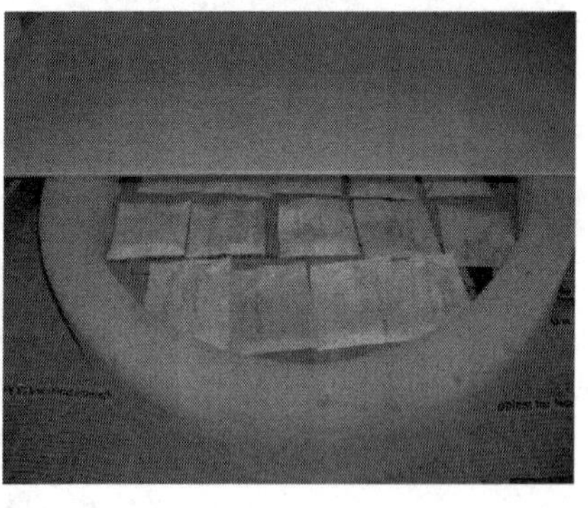

Incubation is a process in which the eggs are made to hatch under an ideal temperature, humidity, light, *etc.* If the process is correctly followed, the rate of hatching and the health of the young worms will be ensured; ultimately the cocoon quality and yield also will be improved. The disease free layings (dfls) of eri silkworms are kept in the egg boxes for hatching.

The dfls are kept in the paper boxes or plastic egg hatching box for uniform hatching. Incubation of eggs at optimum conditions of temperature and humidity is essential for uniform embryonic development and good hatching. Eggs should be incubated in a well-maintained room or incubator at 24-26 °C temperature and 80-85 % relative humidity.

The colour of the egg changes to a dark bluish before two days of hatching. This is the pigmentation stage and is very sensitive. In this stage, the egg should be kept in total darkness or wrapped with black cloth or paper. It indicates that the eggs are ready to hatch. Eggs should be exposed to light in the early morning hours (6 – 8 AM) on the expected day of hatching to get uniform hatching simultaneously.

Eggs generally hatch on 9 to 10th day in summer-autumn seasons and 15 to 20th day in winter. During incubation, the eggs are kept and spread uniformly in a thin layer in the box to facilitate good hatching.

Brushing of worms

Brushing is the transfer of newly hatched larvae from eggs to the rearing bed. The larvae hatch out in the early morning hours and continue up to 9-10 AM. On the first day of hatching, tender leaves are put inside the egg box in the early morning.

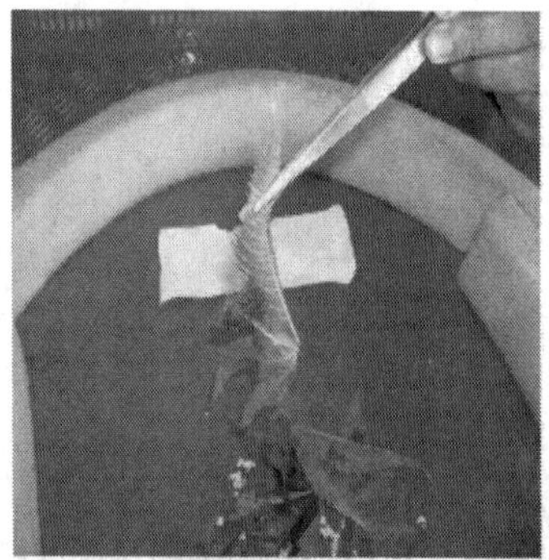

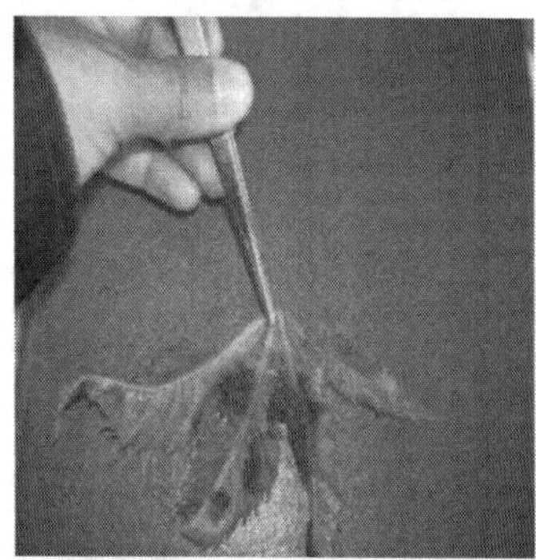

Thenewly hatched larvae crawl onto the leaf and start feeding. The leaves along with the larvae are transferred to the rearing tray with new fresh leaves. The remaining larvae inside the box are gently transferred to the tray with the help of fine and soft brush or bird's feather. The white or creamy colour of the feather or brush is preferred to distinguish the larvae.

The worms hatched in the first two days (48 hours) show more healthy, good vigour and growth; and they are considered for stock maintenance of races. However, the worms brushed in first 3 days are generally considered for commercial rearing. Preferably brushing should be completed during the early hours of the day.

Methods of rearing

There are three methods of eri silkworm rearing, *i.e.*, traditional bunch- rearing, improvcd tray and platform rearing. The farmers of the North Eastern region generally employ bunch rearing method.

Bunch rearing

In bunch rearing method, about 10-12 leaves of castor or branches of kesseru are tied together to make a bundle and hung vertically on a horizontal bamboo/wire/ string support. Then the worms are allowed to feed on the tied leaves. The foliage is changed by keeping fresh bunch near the exhausted one and the worms crawl over the new one. Just below the hanging bunches bamboo mat or tray is kept on the floor so that the worms which fall down are not contaminated with dust on the floor and can be picked-up and put on the bunches. Bunch rearing is simple and easy with minimum cost but yields a better crop due to more hygienic condition. In this method, minimum manpower is utilized for bed cleaning, etc.; but strict maintenance is required like timely replacement of old bunches. Besides, there is no soiling of the leaves due to excreta of the worms as these are fallen down directly beneath the bunch. However, more worms cannot be accommodated on a bunch and more space is required for large scale rearing.

Bunch rearing

Tray rearing

Tray rearing

In tray rearing method, the worms are reared providing the leaves on the tray. Trays are made up of either bamboo or wooden in different shapes and sizes. The shapes are round (bamboo made), square and rectangular (wooden). The young age (I-III instars) silkworm rearing is conducted either in the wooden trays of 50 x 60 x 5 cm size or in bamboo tray (1.0 m diameter). However, bamboo trays of size 1.0 m diameter is more convenient to rear 10-15 dfls until 2nd / 3rd instar; while 600-700 worms can be reared up to 4th instar and 300 worms in the final instar which also provides sufficient space.

Platform rearing technique

This is new innovative rearing method of eri silkworm (Debaraj *et al*., 2003). The model platform rearing device for eri silkworm rearing consists of 3 nos. platforms each of 1 x 2 m size and made up of bamboo strips with sieve size of 1sq cm. Platforms are placed in 3 tier in bamboo rack of size l 2.2 x b 0.75 x h 1.60 m.

Two nos. of such racks can be placed in a room floor area of 5.4 sq m. (1.2 x 4.5 m). Maximum of 1200 eri silkworms at 5th instar can be reared in each platform to accommodate 7200 silkworms by brushing 25- 30 dlfs..

To collect litters of silkworm, gunny cloth is to be fitted below each tier. The technology is found to be advantageous to accommodate almost double quantity of silkworms per unit against the traditional round bamboo try (1m diameter with capacity of 300 nos. 5th instar worms) rearing system. According to quantum of rearing, size of the device may be made.

Silkworm rearing

In general, silkworm rearing divided into two main phases, viz., i) Young age silkworm rearing and ii) Late age silkworm rearing depending upon the nutritional requirements and environmental conditions to be maintained during rearing. In both the phases of rearing, the requirements are different and the techniques of rearing are also quite distinct. The first to third instar worms form the young age and the remaining two instars (fourth and fifth) form the late age worms.

Young age silkworm rearing

- Young age silkworm rearing is also known as "Chawki rearing." A good cocoon crop at the end of successful rearing is greatly influenced by the conditions of young worms rearing. Therefore, young age worms should be reared under good rearing conditions providing them with suitable and good quality leaves.
- Temperature of 26-28 °C and relative humidity of 85-90 % are ideal conditions for young age rearing. Sufficient tender (glossy) and fresh leaves should be

provided to young age larvae. Maximum care should be taken not to expose young age worms to extreme heat or cold. The young age worms are very delicate and should be handled with utmost care. In spite of normal care during young age worms rearing, the chances of loss of worms are more in first and second instar than the grown up worms.

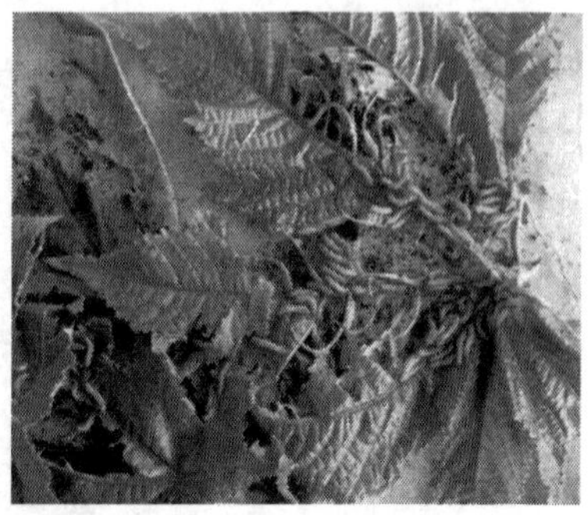

Late age silkworm rearing

Late age worms consist of fourth and fifth instars which need preferably lower temperature and humidity during rearing conditions. These stages consume more quantum of food than the young age worms because the worm has not only to develop silk glands and to increase growth rate, but also has to store up the reserve food materials for the future stages like pupa and imago. Therefore, these stages should be provided as much as quality food they require.

The late age worms consume 80-85 % leaves supplied during the entire larval period. They attain significant growth during this stage. If the chawki rearing is conducted perfectly resulting healthy and robust worms with less mortality, the late age worm rearing will be easy with more chances of a successful crop. But proper care is essential to obtain the full potential of larval growth, maximum yield and best cocoon quality providing with sufficient food.

The environmental and nutritional conditions required in late rearing are different from that of young age. The ideal temperature range 24-26 °C and relative humidity range 70-80 % should be maintained during the rearing of late age worms. These stages should be provided with semi-matured leaves. The feeding of dried, yellow and diseased leaves deteriorates the health of the worms and even death due to diseases. Since 80-85 % of leaves are consumed by these stages, ensure continuous supply of quality leaves after preservation in the leaf chamber.

Maintenance of larvae

During the entire rearing period, the stage wise general maintenance of the larvae is required considering the following aspects,

a) Feeding and its frequency,

b) Bed cleaning,

c) Spacing of worms,

d) Handling of moulting worms and care,

e) Collection and destruction of weak, diseased and undernourished larvae.

Feeding and its frequency

- The suitability of food plant leaves differs according to the period of larval growth. Without physical and biochemical knowledge, the leaf quantity cannot be judged according to the position of the leaves on the plants. While plucking leaf from the same shoot, the softness and the degree of maturity may vary widely according to the position of the leaves. Therefore, it is desirable to pluck more than one leaf from one shoot which appeared to be suitable for the worms' particularly young ages.
- After collection, the leaves should be washed in water and preserved in the leaf preservation chamber covering with wet gunny cloth / bag all around. It is better to provide castor leaves without petiole in tray rearing. 4-5 feedings should be given per day at regular intervals during the young age rearing. In late age worms, 5 feeding per day are essential.
- In the night time more than sufficient leaves should be provided to meet the requirement of leaf consumption throughout the night. It is advisable to prepare a feeding schedule and follow till the end of the rearing strictly.

Bed cleaning

- As soon as the larvae grow-up, the unconsumed leaves and litter increase in the rearing bed which ultimately favour multiplication of pathogenic organisms. Hence, timely bed cleaning is essential to keep the worms healthy.
- Frequent cleaning is better but it involves more labour and ultimately silkworm rearing uneconomical. Only one cleaning is sufficient during first stage worms.
- In second stage, two times bed cleaning is required. In the 3rd and 4th stage three times bed cleaning is required. In the fifth stage, the consumption increases comparatively more than the other instars, ultimately the bed becomes thick and damp soon. Therefore, it is necessary for daily bed cleaning in this stage.
- The method of bed cleaning practiced in eri culture is simple and easy. Prior to bed cleaning, the worms along with the new foliage should be transferred to a new rearing tray carefully. Maximum care should be taken not to harm the worms during handling. Bed cleaning is therefore to be accomplished through experience.

Spacing of worms

The maintenance of optimum number of worms per unit area according to the size or stage of the worms during rearing is called spacing of worms. Proper spacing and good aeration keeps the worms healthier.

- Overcrowding of the worms in the tray leads to competition for food and space and ultimately undernourishment and unhealthy growth of the larvae often results in crop loss. Optimum spacing is, therefore, to be accomplished through experience.
- Spacing of the worms should be maintained along with cleaning. Two times rearing space is sufficient from first stage to third stage. In fourth stage, it may be necessary to increase the space by two or three times and again in the fifth stage, two times space is required to increase than the previous stages. In a standard size rearing tray (1.0 m dia), 300 number of fifth stage worms can be reared conveniently.

Handling of moulting worms and care

- A good rearing is judged by the uniformity of the larvae entering into moulting and emerging from moulting.
- The brushing and feeding of the worms play key role for uniform moulting, if handled properly. When worms enter to moult, stops feeding, became lethargic and less motion. If 75-80 % of the worms enter into the moult, there is no need to feed the rest of the worms. But first feeding should be provided when 80 % of the worms emerge out of moult.
- The worms moult four times during its larval life span. Moulting is a very sensitive period during which the worms cast off its old skin and the body is soft and delicate.
- The larvae take 12-36 hours to complete the moulting process during the different instars and different seasons. It is important to keep the rearing bed dry when the worms are in moult. No bed cleaning should be done during moult.

Collection and destruction of weak, diseased and undernourished larvae

During rearing, the weak, injured, diseased and irregular worms should be collected immediately and put in 2 per cent formalin solution. Such worms should be buried or burnt carefully to prevent spread of diseases. If rearing continues mixing with these worms, the loss due to contamination and disease spread will be more; even crop failure may take place.

Matured worm collection and mounting

- After completion of larval life span, the matured 5th stage larvae discard complete excreta consisting of liquid and semi-solid substances and ready for spinning cocoons.
- Before spinning, the worms stop feeding and become restlessly moving here and there in search of a suitable place for cocooning.
- The matured worms produce a hollow sound when it is rubbed gently between fingers. This is the time for picking the ripe worms and putting them on mountages.

Plastic mountage

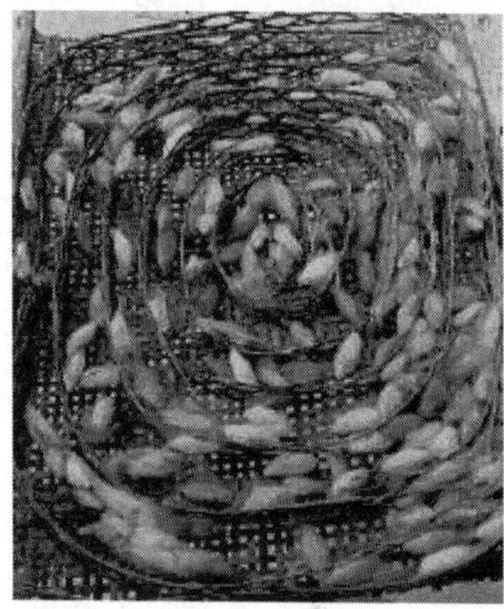

Chandraki

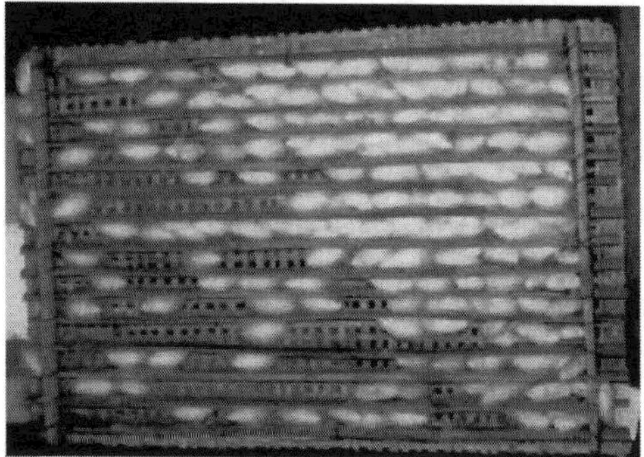

Bamboo strip type mountage

- Before mounting process, the required number of mountages should be kept ready well in time. The worms should be collected carefully for mounting. While cocooning, it is observed that the worms require at least two supporting sides.
- The size of the cocoon depends upon the size of the space available for cocooning. The rearing of the worms takes place during the day time till midday. The commonly used mountages are chandraki, basket filled with dry

leaves, *jali* (a bundle of dry leaves of mango, banana leaves, jack fruit, ornamental plants, etc.) and gunny bag filled with dry leaves. Besided, bamboo chandraki, plastic mountage are also used for cocoon formation.

- Recently, for cocooning of eri silkworm ripened worm a simple bamboo strip type mountage has been fabricated. In this mountage good cocoon recovery is 98.9% against 97.43% in conventional jail system and inferior cocoon 1.09% as compared to 2.56% in Jali. Shell weight 0.52g as compared to 0.40 g in jail (Debaraj, *et al.*, 2012).
- The same has been recently innovated as wooden strip type mountage, which is collapsible in nature. Harvesting of 300 cocoons can be done in 3 minutes in this mountage against 30 minutes in traditional *jali*.

The leaves should not be completely dried as semi- dried leaves are suitable for easy spinning. After keeping the optimum number of worms in the respective mountages, it is covered by newspaper or cloth to make support and semi- dark, a suitable condition for cocooning. However, if disturbed, it stops spinning for a short period.

Due to unavailability of suitable place and disturbance, the larvae spin defective cocoon and even fail to spin cocoons. In some cases, the physiological condition inside the body is abnormal and the larvae often die without pupating.

Mounting of immature and over matured larvae results in poor cocoon quality. The quality of cocoon also depends upon the type of mountages, density of worms in mounting and different mounting methods / models.

During spinning, temperature, relative humidity and aeration influence cocoon quality. The ideal condition for spinning is around 24 - 25 °C temperature and 60-70 % relative humidity. While mounting, the optimum number of worms should be maintained per mountage, i.e., 300 worms per chandraki of 1.0 m diameter size.

Harvesting and assessment of cocoons

After completion of spinning, the larval skin is cast off and pupation takes place. The pupa has a thin cuticular skin which is soft and may get ruptured easily, if disturbed. The last and important step is harvesting of cocoons from the mountage in time. Cocoons should be harvested after 5-6 days of spinning in summer and 8-9 days in winter. The harvesting process is the best time of sorting of cocoons according to the quality. The cocoons should be sorted out as good, double, melted, stained, dead, inferior, cut or pierced cocoons. Good commercial cocoons should be shifted and dried perfectly after the harvest. Cocoons should be preserved carefully to protect from fungal infection and attack from pest and predators. Cocoons should be assessed on the basis of cocoon weight, shell weight and silk ratio.

PROCESSING OF ERI COCOONS

Processing of eri cocoons

The eri cocoons are open mouthed since its silk filament is discontinuous. Hence, eri cocoon can only be used for spinning purpose. Eri silk has certain excellent textile properties such as fineness (2-2.5 denier) and thermal properties which play important role for determining the end use of a fibre. Eri silk is finer than muga and tasar silk. Major portion of eri cocoons produced in the region is locally spun through traditional devices like Takli and other spinning devices like CSTRI spinning wheels etc.

Drying

Sun drying is usually practised because of its simplicity. However, hot air drying is preferable where cocoons are kept in 95 to 55 0C for 3-4 hours. As eri pupa is mostly consumed by the people in this region, stifling process is not required. However, shell drying is necessary for preservation and storage.

Cocoon selection: Clean, dry and uniform quality cocoons are to be taken for spinning.

Degumming

Traditional process

The cocoons are loosely tied in cotton cloth and boiled in 10g Soda/l of water for 45 minutes to 1 hour. After boiling, individual cocoons are stretched or opened up in plain water into thin sheets. 3-4 such sheets are joined to make a cake, which is dried and used for spinning in Takli. Locally available materials such as ash obtained from banana, wheat stalk, paddy straw and pieces of green papaya are commonly used as degumming chemical instead of soda.

Improved method

Eri cocoons are loosely tied in a porous cloth and the bundle is immersed in an alkaline solution of 10-12g soap and 2-4g Soda per litre of water and bolied for an hour. The cocoons are then washed and reboiled in fresh water for 15-30 minutes. After proper washing the cocoon shells are dried without disturbing the fibre layer and then utilized for spinning especially in CSTRI machines.

Spinning

Takli spinning

The takli consists of a spindle with disc like base. The spinner holds the cocoon cake in the left hand, drafts and then feeds the strand with the right hand to the spindle. The spindle is occasionally rotated by the right hand in order to wind the yarn to the spindle. Production is around 40-60g/ person/days.

Improved spinning wheel

Although the Takli is very simple and cheap, its output is quite low. Improved spinning devices have been developed time to time in which CSTRI spinning wheel is the latest one. The production is around 120-150g/ person/ day with 70-80% recovery from the cocoon shell.

Weaving

Generally throw shuttle looms are conveniently used for weaving eri cloth. The preliminary operation for weaving includes sizing and warping which is winding of the thread for warp. The processes are mainly manual. The warp is prepared section by section either in horizontal drum or in hand reel. The fabric can thus have the required width. The weft thread is fitted on a bobbin into a boat shaped shuttle. The finished fabric is wound on the cloth beam steadily.

Now a days, fly shuttle loom is used for better equality eri fabric and also for blended fabric. The production quantum becomes 2.5 times higher than the throw shuttle loom. A throw shuttle loom can weave about half 0.5m cloth/ day working whereas the fly shuttle can weave upto 5m.

PESTS AND DISEASES OF ERI SILKWORM (*SAMIA RICINI*)

Pests

1. Uzifly - *Exorista bombycis*. F. Tachinidae. O. Diptera

Morphology of Uzi fly adults

§ Adults are blackish grey in colour.

§ The head **triangular** in shape.

§ On the dorsal side of thorax, there are four longitudinal black bands.

§ The abdomen is **conical.**

§ The first abdominal segment is **black** and the rest, **greyish yellow**.

Morphological differences between male and female Uzi fly adults

Sl. No.	Character	Male	Female
1	Body length	Longer (12 mm)	Relatively shorter (10 mm)
2	External genitalia	Covered with brownish orange hairs on the ventral side of the abdominal tip	Not so.
3	Lateral regions of abdomen	Covered with bristles which are more dense	Bristles not dense, restricted mostly to last two segments
4	Longitudinal lines on the dorsum of the thorax	More vivid	Less vivid
5	Pulvilli	Larger	Smaller
6	Life span	10 - 18 days	2 -3 days longer than males

Type of damage

§ Mature maggot causes reduction in yield of cocoons and cocoon quality.

§ Causes death of silkworm larva.

Symptoms of damage

§ Presence of creamy **white oval** eggs on the skin of larvae in the initial stage.

§ Presence of **black** scar on the larval skin

§ Silkworm larvae die before they reach the spinning stage (if they are attacked in the early stage).

§ In later stage, **pierced cocoon** is noticed.

Period of occurrence

§ Throughout the year, severity is more in **winter months**

Life cycle

- Gravid female lays eggs **(300 eggs) @ 1-2 eggs on each silkworm.** Eggs hatch in **2-3 days**. Young maggots pierce into the host body.
- A small black scar develops on the cuticle of the host.
- The **maggots has three instars.** First two instars suck **the body fluid of larva and the third instar devour the internal tissues**.
- The maggot after devouring 5 – 8 days of parasitic life, emerged from the host body and kills the host in that process.
- The mature maggot remains for 12 – 20h post feeding and pupate in loose soil or in the crevices on the rearing floor.
- Pupal period is 10 – 12 days. Life cycle is completed in 18 – 24 days.
- **The damage by uzi fly is not much significant to eri culture as the cocoons are not reeled.**

2. Lizards

- Feeds on the larva during the rearing period.

3. Ants

Manomorium indicum ,Camponotus compressus

- The ants attack cocoon, pupae, adults and eggs in grinages.

DISEASES OF ERI SILKWORM AND THEIR MANAGEMENT

The eri silkworms are affected by **pebrine, flacherie and grasserie.** Hygeinic rearing conditions and provision of good quality leaves prevent infection by the pathogens.

1. Pebrine- Protozoan disease

Caused by *Nosema* sp. The pebrine spores from *Bombyx mori* is not infective to eri worms and vice versa. The spores are 3.2 - 4.6 microns in length and 1.5 - 2.1 microns in width. The disease is sporadic in nature. Transmission is both trans ovarial and oral. The incidence of the disease is sporadic.

Symptoms of disease

Uneven size of the worms in the rearing tray is the characteristic symptoms.

- Sluggish nature of worms
- Reduced feeding
- Body of affected worms becomes thin and darker
- Stunted growth
- Black pepper like spots over the integument. Appear after only second moult.
- Moth emergence is affected and the emerged moths are with small, deformed and scorched wings. The scales are thin in density and adhere loosely.
- Irregular laying of eggs seen
- Reduced fecundity
- The eggs laid by infected moths do not hatch.

2. Muscardine

Causal Organism: It is a fungal disease caused by *Botrytis bassiana*

Occurrence: This occurs in all the crop seasons due to dampening weather.

Factors responsible for occurrence of disease: Rearing in damp, shady place and foggy conditions in winter and improper bed cleaning sometimes lead to the occurrence of this fungal disease.

Damage: Damage is due to loss of vitality and death of silkworms.

Symptoms

- The infected larva becomes sluggish, swollen and entire body is covered with fungal spores and gets mummified at later stage.
- The disease generally occurs during winter, infecting the late age and mature larvae.
- The affected worms limp, looses elasticity, cease to move and die rapidly.

Control Measures

- Select well-aerated rearing rooms having sufficient sunlight during winter.
- Daily bed cleaning during fourth and fifth stages followed by dusting of Carbendazim (1%) or Slacked lime and Bleaching powder mixture at 9:1 ratio.
- Burning of sulphur for sulphur fumes in the rearing house is effective.
- Avoid feeding tender leaves to late age worms.
- Pick out the weak, irregular and diseased worms and destroy them.

3. Flacherie- Bacterial disease

- In eri silkworm, the disease is caused by **virus followed by bacterial attack**.
- Loss due to the disease is **14-30 %.**

Causal Organism

- Flacherie is first caused by virus followed by bacterial attack.

Occurrence: The disease prevails all through the year, but the incidence is more prevalent during summer and rainy seasons.

Factors responsible for the disease

- High temperature and humidity, poor ventilation, poor sanitation, accumulation of faeces in trays and feeding of contaminated leaves are the predisposing factors.

Symptoms

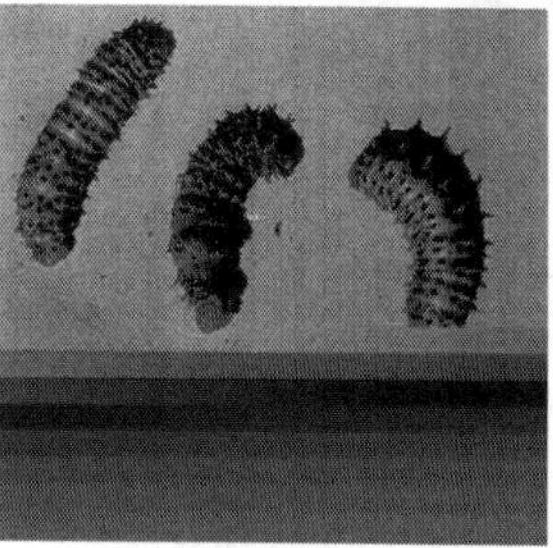

- The infected worm loooses clasping power and starts vomiting.
- Loss of appetite, sluggish movement, swollen thorax, shrinkage of abdominal segments incomplete moulting, and soft and sticky excreta are also observed.
- Body blackens and rots at later stage.
- The disease generally occurs in late age worms.

Damage

- Loss due to this disease accounts for 14-30 per cent.

Control Measures

- Sprinkling of Bleaching powder and Slacked lime mixture at 1:9 ratio on the rearing beds immediately after bed cleaning as well as on the floor @100 g/ sq m area with proper disinfection of rearing room and appliances.

Precautionary Measures

- Avoid overcrowding of worms, maintain optimum temperature (26-28° C) and RH (80-85%) in rearing house, avoid feeding of diseased/contaminated leaves to worms, pick up diseased worms and burn or bury them and avoid feeding dry and dirty leaves during late age rearing.

4. Viral diseases of eri silkworm

- ***Samia ricini* is affected by both cytoplasmic and nuclear polyhedrossi virus.**
- ***Samia ricini* is immune to grasserie but it is reported in *Philosamia cynthia*.**

Symptoms

- Swollen body
- Restlessness of larva
- Intersegmental swelling
- Shiny appearance of worms.
- **Wipfelkranheit or tree top disease.** This is the expression of **hanging symptoms.** The worms move to the periphrey of the rearing trays and hang upside down.
- Skin fragile and Oozing out of white fluid.

Maternal inheritance of grasserie is reported in *P. ricini* and *P. cynthia* but the cross (F_1) *Samia cynthia* is resistant. F_2 were found susceptible to grasserie.

i. Cytoplasmic polyhedrosis virus

Symptons of disease

- Vomoting and diarhoea
- Reduced feeding, sluggishness of worms
- **Rectal protrusion** is the characeristic symptom sof CPV
- **Milky appearance** of mid gut.
- **Causal Organism:**It is a viral disease.

Occurrence

- Occurs during the rainy and summer seasons when there is excess of moisture in leaves.

Factors responsible for occurrence of disease

- Fluctuation in temperature and humidity, feeding of tender and contaminated leaves, improper bed cleaning and poor ventilation are the predisposing factors.

Control Measures

- Sprinkle bleaching powder and slaked lime mixture at 1:9 ratio on rearing bed immediately after bed cleaning.
- Rearing worms under proper sanitation and hygiene. Avoid overcrowding of worms in rearing bed.
- Avoid feeding tender leaves during late age.
- Clean the rearing bed every day during fourth and fifth stages.

MANAGEMENT OF PEST AND DISEASES OF ERI SILKWORM

Management Practices of uzi fly, *Exorista bombycis*

a. Cultural and mechanical control

- Collect and destroy all the uzi pierced larvae from the rearing tray, crawling maggots falling from mountage or rearing tray and pupae.
- Sort out all the uzi pierced cocoon and burn it along with the cocoon wastes.
- Keeping the rearing room floor free from cracks and crevices to avoid pupation of the maggots.
- Provide uzi proof wire mesh or nylon net to the windows, ventilators and doors of the rearing room, in order to prevent the entry of uzifly adult.
- Periodical checking of uzi proof net for any holes.
- Provide nylon net to the rearing trays/ shoot rearing racks to prevent egg laying of gravid uzifly female.
- Provide ante-chamber in front of the rearing room door as uzifly proof.
- Collective skipping of atleast one batch of rearing by the farmers in endemic areas prevents multiplication of uzifly due to non- availability of host.
- Sorting out all the uzi pierced cocoon before transportation to cocoon market to check further spread of uzifly.
- Care must be taken while taking the harvested leaves inside the rearing house.
- Transport cocoons in cloth bags.

- Disinfect / immerse the cloth bags in pesticide solution after disposing cocoons in market.

b. Chemical control

- **Setting up of uzi trap** : **It is a chemo trap available in tablet form. Dissolve two tablets per uzi trap in one litre of water or mix Asiphor @ 17 ml / litre.** This solution should be kept in white colour basin (avoid using metal) and kept near the windows and doors, both inside and outside of the rearing room, possibly near light sources. It can be used from the 3rd instar onwards. The yellow colour solution will attract and kill the adults, which fall into it. Change the solution once in three days.
- **Uzicide spray** : It is an **ovicidal chemical(1.0 % benzoic acid)**available in liquid formulation. It has to be sprayed on the silkworm larval body starting from 3rd instar onwards. Spraying should be taken on 2nd day of third instar, 2nd and 4th day of fourth instar and 2nd, 4th and 6th day of fifth instar. The spray fluid required for 100 dfls is one litre for third instar, 1.5 litre for fourth and 2.5 l for fifth instar for each spray. Care should be taken to provide proper ventilation while spraying to avoid excess moisture/ humidity inside the rearing room. Feeding should be given only half an hour after spraying.

c. Biological control

- Releasing of hyperparasitoid: *Nesolynx thymus* an endopupal gregarious parasitoid on uzifly is released into the rearing room @ **one lakh per 100dfls**. Hyperparasitoid should be released in three split doses, *i.e*, **8,000; 16,000 and 76,000** during fourth instar, fifth instar and at the time of cocoon harvesting / 100 dfls, respectively. These releases should be done only in the evening hours and after release the door and window should remain closed.
- The released hyperparasitoid will parasitise the hidden pupal stage of the uzifly and hinders the multiplication.
- *Nesolynx thymus* is available in pouches. **One pouch contains 10,000 numbers costing Rs. 10 / pouch.** The pouch should be tied in thread and hanged inside the rearing room for easy emergence. Do not release the parasitoid inside the rearing room on the day of disinfection with bleaching powder.
- **Sequential release of hyperparasitoids *viz., Nesolynx thymus* and *Exoristobia philippinensis* / *Trichopria* should be done especially during summer months.**

Management of silkworm diseases

Maintenance of hygiene

- Sun drying of rearing appliances.
- Regular disinfection of rearing room and appliances with 5% bleaching powder
- Ensure proper ventilation for aeration and maintenance of optimum temp and humidity and aeration.
- Disinfection of rearing room, trays and discarding of diseased worms.

Healthy rearing practices or prophylactic measures to be followed for preventing diseases

- Provide adequate bed spacing
- Provide nutritious mulberry leaves
- Collect infected larvae, faecal matter and bed refuse and burn
- Early diagnosis and rejection of infected lots
- Keep the rearing bed thin and dry at the time of moulting.
- Apply bed disinfectants like Vijetha / RKO / Sakthi Seri Dust @4 kg / 100 dfls on the larva after each moult ½ hour before resumption of feeding.
- Avoid spraying commercial BT insecticides nearby mulberry field.
- Surface-disinfect the layings in 2% formalin for 10 minutes before incubation.
- Proper disinfection of rearing room and appliances.
- Proper disposal of diseased larva in soultion containing 2 % bleaching powder and 0.3 % slaked lime .

Disinfection of eri rearing room

- Wash rearing room and appliances with 5% bleaching powder solution.
- Keep the appliances inside the rearing room and seal the room.
- Fumigate 5% Formaldehyde solution under high humid condition or with 2.5 % chlorine di oxide.
- Open the room after 24 hours.
- Disinfect the rearing room at least 3 days before and soon after rearing.
- In case of perbrine incidence the rearing house must be disinfected with 2% formaldehyde.

Management practices

Viral diseases

- Proper disinfection with the right type of disinfectant at the right dose.
- Maintaining proper temperature and humidity conditions.
- Avoid feeding of young leaves or tender leaves to older larva. Clip off the tender leaves at the tip of shoot in case of shoot rearing.
- Avoid long storage of harvested leaves or shoot.
- Avoid starvation
- Disinfection of rearing bed with bed disinfectants at the rate 4 kgs/ 100 dfls.
- Selection of breeds according to season.
- Dip the mulberry leaves in 1.0**% aqueous or 800 ppm hexane leaf extract of *Psoralea corylifolia*, shade dry and feed to the worms after second, third and fourth moult.**

Bacterial diseases

- Leaf quality is an important aspect to be considered in case of bacterial flacherie.
- Avoid feeding of nutrient deficient leaf, over matured leaf, dried leaves, leaves collected from the shady areas, soiled and dusty leaves.
- Avoid any damage to the skin of the worms.
- The leaves collected from garden, nearby road should be washed and shade dried before feeding.
- Proper disposal of bed waste and affected larva to prevent the spread of the disease.
- Avoid overcrowding of worms.
- Avoid fluctuations in temperature and humidity conditions.
- Avoid spraying of *Bacillus thuringiensis* for pest control in the fields nearby mulberry garden. If so a safe distance of 100 m should be given.
- **Feeding of worms with mulberry leaves treated with 0.05 % Streptomycin sulphate, chloramphenicol or erythromycin.**
- Bed disinfection with bed disinfectants at 4 kgs/100 dfls.
- Feeding of worms with leaves treated with 1 **% *Thuja orientalis* aqueous leaf extrac**t.

Muscardine or fungal disease.

- Proper disposal of infected worms is essential as the conidia formed from mycelium of infected cadaver spread more easily in air and result in spread of the disease
- Disinfect the rearing house with 2 % bleaching powder solution containing 0.3 % lime
- Dust lime at the rate of 3 g/ sq. feet on the rearing bed when needed to reduce the humidity of rearing bed and at the time of moulting.
- The bed disinfectant, Suraksha is specific for muscardine disease.
- Dust 1 % Dithane M 45 (10 g of Dithane M 45 per kg of lime) at the rate of 3 g/ sq. feet on the first and second instar worms and 2 % Dithane M 45 on third and fourth instar worms @ 5 g/ sq. feet
- As there is a possibility of spread of white muscardine through mulberry leaf webber , it is highly essential to manage the leaf webber.

Pebrine/ Protozoan disease

Prevention and control of pebrine

- Production of healthy eggs (through mother moth examination)
- Destroy dead eggs, dead larvae in the cocoon, dead moths, excrement of larvae from infected trays, exuviae of infected larvae and other possibilities of infection such as contaminated litter.
- Effective disinfection of rearing house
- Disinfection of rearing appliances with hot water.
- Maintain hygienic conditions during rearing.
- Surface sterilize the layings/ eggs with 2% formalin for 10 minutes before incubation.
- Inspection of egg shell and larvac
- Immediate destruction of infected crop.
- Monitoring and examination of unequal sized larvae in the rearing bed.
- Seed legislation in the true sense needs to be implemented.
- Mother moth examination.
- The silk worm bed waste should not be applied to mulberry field.

Management of pebrine

1. **Physical method of pebrine control**
   - Exposing 1.5-2.5 days old eggs to water treatment at 46°C for 90-150 min or at 48^0C for 50-70 min.
   - Exposing the pupae to 40^0C for 8 hours. (Middle / late stage pupa) .
2. **Chemical method**
   - Asiphor 2% (alkyl phenoxy poly glycol) disinfection for 24 hours controls all pathogen
   - Benomyl and Mebandazol are common anti-protozoan drugs control the disease.

a. **Mass mother moth examination:** In the Commercial Grainages, a system of mass moth examination is adopted.

1. Homogenize 20 egg laid fresh moths or dry moths (dried at 70^0C for 6 h)] for 1-2 minutes, in a domestic mixier by adding 80 ml (fresh moth) to 90 ml (dry moth) of 0.6 % K_2CO_3 solution.
2. The homogenate is poured into a clean beaker and allowed to stand for 3 -5 minutes.
3. Filter the homogenate by using cotton or muslin cloth. Volume of filtrate recovered will be around 70 ml.
4. Transfer the filtrate into 100 ml capacity centrifuge tubes.
5. The filtrate is centrifuged at 3000 rpm/3500 rpm for 3 min / 5 min.
6. The supernatant is discarded and the sediment is dissolved in few drops of 0.6 % K_2CO_3 solution.
7. The dispensed sediment is examined under microscope at 600 x.
8. Two smear should be examined per sample. Minimum of five microscopic fields are examined per smear. Two persons should check each smear (Refuse should be discarded into a pit after disinfection with 5 % bleaching powder)

b. **Individual mother moth test / Pasteur Test:**

This method is a best method advocated for basic seed production and 100 % seed produced should be subjected to examination.

Here, individual moths are crushed thoroughly in a medium of 0.6 % K_2CO_3 solution. The rest of the procedures are the same except for the use of smaller tubes for centrifugation.

FOOD PLANTS OF TASAR SILKWORM AND MANAGEMENT

Type of the tasar silkworm	Scientific Name	Host plants	States in which reared
Tropical tasar	*Antheraeamylitta*	*Terminalia tomentosa* (asan) *T. arjuna* (Arjun) *Shorearobusta* (soal) *Zizyphusjujuba* (ber)	Bihar, Orissa, Madhya Pradesh, Uttar Pradesh, Andhra Pradesh, Maharashtra, Karnataka
Temperate tasar	*A. pernyi* *A. proylei* *A. yamamai*		Jammu and Kashmir, Arunachal Pradesh, Mizoram

Tasar silkworm is wild and reared outdoor. Tropical and temperate tasar silkworms are reared in India. Tasar worms are uni/bi/trivoltine. It undergoes pupal diapause.

Primary food plants

Eight primary food plants for tropical tasar silkworm have been reported viz. *Terminalia tomentosa* W. & A. (Asan), *Terminalia arjuna* W. & A. (Arjun), *Shorea robusta* Roxb. (Sal), *Lagerstroemia parviflora* Roxb. (Sidha), *Lagerstroemia speciosa* Pers. (Jarul), *Lagerstroemia indica* Linn. (Saoni), *Zizyphus mauritiana* Lam. (Ber) and *Hardwickia binate* Roxb. (Anjan).

Secondary food Plants

The more than two dozen secondary food plants so far reported, the most important are *Terminalia chebula* Retz.(Haritaki), *Terminalia belerica* Gaertn. (Bahera), *Terminalia catappa* L. (Janglibadam), *Terminalia paniculata* Roth. (Kinjal), *Anogeissus latifolia* Wall. (Dhaunta), *Syzygium cumini* (L.) Skeels (Jamun), *Careyaar borea* Roxb. (Kumbi) and *shorea tailura* Roxb.

Distribution: The food plants of tropical tasar silkworms grow luxuriantly at low altitude (0-600 MSL) between 40° north and south latitudes.

Terminalia arjuna food plant cultivation

I. Introduction

The botanical name for Arjun is *Terminalia arjuna* and it belongs to the family *Combretaceae*. The local names or the vernacular names for Arjun are given as: (1) Assamese - Orjuno, (2) Bengali -Arjhan, kahu, (3) Gujarati -Arjuna, Dhulasadar, Arjun sadada, (4) Hindi -Anjan, Anjani, Arjan, Arjuna, Jamla, Kahua, Koha, (5) Kannada -Holematti, Maddi, Torimatti, (6) Malayalam -Attunarathu, Vellurnaranthu, (7) Panjabi -Arjan, Jamla, (8) Marathi -Anjan, Arjan, Azun, Sasura, Sanmadat, Savimadat, (9) Tamil -Kulamaruthu, Tanikai, Veldati and (10) Telugu -Erramaddi, Yeramaddi.

Arjun is a large, handsome evergreen tree of tropical and sub-tropical forests. In its natural habitat, the tree occurs along the banks of rivers, and streams and is common in most part of the country attaining a height of about 25 m and girth of about 3m.

The occurrence of the tree in natural forests as food trees has helped in developing the tasar production as forest based industries in the states of Bihar, Orissa, Madhya Pradesh. This is also one of the important tree under social forestry programme for plantation in afforestation. The tree is extensively planted for shade or ornamental purposes. Some parts of the tree have got medicinal importance.

Terminalia arjuna grows throughout the greater part of India chiefly along rivers, streams, ravines, and dry water course. It is quite common in Chota Nagpur, central India, and in some parts of Madhya Pradesh and Tamil Nadu. It extends I northward to the sub-himalayan tract, where it grows sporadically along river banks. It is a characteristic tree of dry tropical riverain forests and riparian fringing forests, and is one of the predominant species in Gir forests of Saurashtra.

II. Description

Arjun is an evergreen tree with spreading crown and drooping branches. The stems are rarely long or straight, generally always buttressed. The bark is thick, smooth, exfoliating in large irregular sheets. The leaves are sub-opposite, coriaceous, usually 10-15 cm long, cordate, shortly acute or obtuse at apex. Flowers are in panicled spike. Fruits are 2.5-5.0 cm long and ovoid with 5-7 hard winged angles.

III. Environmental Requirements

The environmental conditions required for suitable growth of Arjun tree includes rainfall, temperature and soils.

1. **Rainfall:** The plant is found in its natural range of distribution between rainfall ranging from 750-1900 mm and even as high as 3800 mm.
2. **Temperature:** The absolute maximum shade temperature is between 38-48°C and minimum 0-15°C.
3. **Soils:** It prefers loose, moist, fertile alluvial loam soil along river banks. It is not suitable for dry localities and exposed hillsides. The root system is superficial, radically spreading along stream banks. Cool moist location along streams, ravines, or dry water courses are the ideal places. The soil is enriched because of calcium in the trees.

IV. Life Cycle

The tree bears flowers from April to July, May to July in Central India and the fruits ripen the following February to May. The tree is nearly evergreen and the new leaves appear in the hot season before all the old leaves fall. The tree starts bearing fruits at an early age of about 6-7 years. Every third year is a good seed year.

The fruit requires good soaking for germination. Germination is epigeous, and growth of the seedlings is fairly rapid. During the first few years, the seedlings generally exhibit a straggling habit and develop long side branches. The plants raised in the nursery, however, do not show this character. The plants under natural conditions die back every year for 6-7 years and become bushy. The development of the root is faster, than that of the shoot, the root penetrates about 30 cm in two months after germination. Under favourable conditions, the seedlings attain a height of about 45 cm in one year and 2-3 m in three years. The seedlings are sensitive to frost as well as to drought.

The growth in the saplings and pole stages is also fairly fast. The poles develop strong branches and consequently a broad crown. A full-grown tree *T. arjuna* is a large tree. The crown is broad with drooping branchlets. Bark is smooth and exfoliating in irregular sheets.

V. Nursery Practices

About 175-450 fruits weigh a kilogram. The tree is nearly evergreen but new leaves appear in the hot season before all the old leaves have fallen.

A. **Nursery site:** Selection of ideal and favourable site is most important factor for raising healthy seedlings. Arjun is a water loving plant and as such nursery should be prepared preferably near regular water resources, along streams or nala sites where necessary facilities of irrigation may be made available. Thorough soil working should be done before seed sowing season. Selection of site should be done where some partial shade is also available and the whole area should be cleaned before the preparation of beds and process of seed sowing. Undesirable weeds and perennial bushes be removed and burnt. Site should be selected near main road or connecting roads for easy transport facilities. Proper fencing of the area is also necessary for protection of nursery.

B. **Nursery techniques:** The Central Tasar Research Station, Ranchi (Bihar) has evolved as suitable technology for raising arjuna plants in large scale by which the germination period is reduced to hardly 7-21 days as against norma! germination period which continues for 2 months. The standardization of nursery techniques include, seed selection, seed viability, seed treatment rooting media and transportation.

   1. **Seed collection:** Optimum time for mature seed collection found to be April-May which also facilitated easy transplantation of well grown

seedlings after three months during the onset of monsoon. Based on variability in weight of seeds for rate were studied for their germination performance.

2. **Seed viability:** The single seeded fruit of arjuna has hard pericarp containing alkaloids, hardness may result in delayed and uniform germination.

3. **Seed treatment:** Soaking in hot water, for 72-96 hours is significantly superior over other treatments. 90% seeds germinate with minimum initial period of germination of 8 days. Boiling of seeds and prolonged soaking has good effect on germination. Hot water treatment is also effective to control seed borne diseases.

4. **Rooting media:** Treated seeds are scarified and sown in polythene bags with combination of rooting media -soil, sand, compost, and mixture of compost., soil and sand in ratio of 3:2:1. The studies with different rooting media revealed superiority of compost mixture of FYM: soil: sand: 3:2:1. The scarified seeds are heaped under tree canopy shade covered with moist gunny bags, watered by sprinkles to maintain moisture. This practice ensures a homogeneous growth of seedlings.

5. **Transplantation:** Seedlings raised in beds and polythene bags are transplanted after three months in pits of 30 cm x 30 cm x 30 cm in plantation during the monsoon. Transplantation of seedlings raised in polythene tubes has been observed to be beneficial with higher survival rate of transplantation.

C. **Seed germinability and seedling growth:** The fruit requires good soaking. Natural seedlings for the first four years exhibit a straggling habit, while nursery raised seedlings do not exhibit this character. Root development is faster than the shoot, the root penetrates 30 cm in two months after germination. Under favourable conditions 45 cm height of seedling is obtained in one year, and 2-3 m in 3 years. The seedlings are sensitive to frost and drought sapling and pole stage growth is fast.

i. **Pricking out:** For speedy growth of the plants, seedlings are picked out from beds into polythene bags or other containers just after the appearance of the first pair of leaves.

ii. **Irrigation:** Regular irrigation of beds and polythene bags is to be done after sowing of seeds and thereafter with intervals for the normal growth of seedlings, especially during hot season.

ii. **Weeding:** Regular and frequent weedings of young seedlings is highly beneficial for their faster growth particularly the areas where perennial grasses come up very frequently. A schedule of weeding in nursery stage is necessary. Plants in plantations may also be weeded one or two seasons till their establishment.

D. **Vegetative propagation/Cloning techniques:** Arjun can be raised successfully by direct seed sowing, by entire transplantor by stump plantings. Its propagation has also been tried by air layering.

   i. **Air layering:** Propagation by air layering is normally applied in plants which do not respond easily through cuttings. Air layering has been successfully tried in arjun tree.

   The age of rooted layers on the mother plant has significant role in the survival. One-month old rooted layers show better survival after transplantation.

   In another study, air layering on four-year-old plants showed rooting after 25 days in month of July. The rooted layers are transplanted separately in the interval of 15, 30 and 45 days and transplanted in pits 60 cm x 60 cm x 60 cm. Sprouting of lateral buds were observed after 20 days of transplantation.

   The comparative effects of Indole acetic acid, Indole butyric acid, and NAA on rooting and survival of air layers of *T. arjuna* has shown that the percentage of rooting in all three hormones showed increasing tendency from lower concentrations to optimum concentration after which it declined sharply.

VI. Planting Practices

The planting practices include several aspects which are mentioned here.

A. **Planting site and its preparation:** The plant does not tolerate dense shade and is sensitive to drought and frost, therefore, planting site should be cleared. Pits of 45 cm^3 size should be dug up at spacing of 3 m x 3 m in the field for planting. Pits should be filled up with a mixture of farmyard manure and sandy loam soil. Irrigation facilities should also be available near planting site. Thorough soil working of the site should be done before sowing operation.

   Arjun has not been raised in regular blocks or mixed plantations. The plants are raised by direct sowings and by planting the nursery raised seedlings.

   1. **Direct sowing:** The tree is raised successfully by direct sowing of seeds. Seeds are sown with the commencement of monsoon rains in June-July in dug-up lines 2-3 m to 4 m. Germination starts in about 20 days and completes in about 7-8 weeks. Early growth of plants in nursery

is rapid. Plants raised by direct sowing can tolerate hot weather better. Seedlings are sensitive to drought and frost. Therefore, periodical weeding is beneficial during rains.

2. **Planting out entire seedlings:** Seedlings are raised in nursery beds or in polythene bags. The nursery raised two to three months old entire plants with ball of earths having average length of root and shoot about 30 cm and 13 cm respectively are planted out removing all leaves except the top pair of leaves. During the early monsoon rains show a survival of 58% and height of 99 cm after three years.
3. **Stump planting:** Arjun trees can be raised successfully by means of stump planting. Suitable stumps are prepared from 12-15 months old seedlings their normal size is 1.3 cm to 2.5 cm collar diameter. Best time for stump planting is the end of July, the seedlings withstand transplanting well during the first rains before the tap root becomes too long. Older seedlings with longer roots become too long for planting are better made into stumps.
4. **Spacing:** Spacing of plants should be 3 m x 3m depending on availability of space and planting material. The plants can also be successfully raised along with other field crops,
5. **Irrigation:** This is very essential requirement for germination, growth and development of seeds and plants, The sown beds and polythene bags or plantation require regular irrigation during summer. Seasonal irrigation schedule should be arranged throughout the year as per necessity. During extreme hot month May to June, watering should be done early morning and later evening hours,

VII. Cultural Operations and its Calender

Nurseries as well as plantation areas should be well maintained with proper cleaning operations. Rotational and periodical weeding schedule should be applied weekly or fortnightly in nurseries and for onc or two years in young plantations. Soil working helps in loosening the intact hard soil for aeration, watering and promotes the healthier growth of seedlings and young plants. A tentative irrigation schedule is highly necessary and very essential for establishment of nursery seedlings. Watering is even more needed during very hot season for the up-keep and proper growth of seedlings in nurseries.

Young seedlings are also liable to get infected and damaged by many organisms, pathogens and insect pests. It is also important to protect the area/site against browsing and destruction from cattles, goats, etc.

HOST PLANT OF *TERMINALIA* PESTS AND DISEASES AND THEIR MANAGEMENT

Out of nearly 18 insect pests damaging *Terminalia arjuna* in different parts of India, only a few are the major pests. A brief account on them is given here.

A. Insect pests

Important pests of Arjun are decribed here.

1. ***Trioza fletcheri:*** This sap sucking insect damage the tree by forming galls on leaves in nurseries, plantation and natural forests. Eggs are laid singly or in clusters on the leaves. The nymphs feed on the sap on the upper surface of leaves causing the formation of small, unilocular galls which open by slits densely clothed with white hairs. It is of little importance in large trees but injury in nurseries and young plantations is sometimes severe.

Control measures

(i) Insecticides like metasystox 0.02%, dimecron 0.02% in water may be sprayed which kill the nymphs inside the gall.

(ii) Spraying of fenvalerate 0.02% to 0.03% as prophylactic spray or monocrotophos 0.03% twice in March-April at 15 days interval which has been proved to be best.

2. ***Antheraea paphia:*** The larvae of this moth are polyphagous. The female moth of this insect are collected for food by many tribes and the cocoons are collected for extraction of tasar silk.

Control measures: Generally there are no records of epidemic defoliation. Moreover the population of the defoliation is kept below economic injury level and the larvae and pupae of this pest are often collected by the tribals and other villagers for tassar cocoon.

3. ***Lymantria mathura:*** It is a polyphagous insect, and feeds on the flowers and foliage. It has two generations in a year.

Control measures

- Foliar spraying of folithion 0.2% or Fenvalerate 0.01% on young trees is also effective.
- The insects which attack the felled logs of arjuna are the beetle of *Lychus africanus* and *Sinoxylon anale.* Preservative treatment by dipping the logs in 1 % sodium pentachlorophenate or boric acid and borax in ratio of 5:1:1 in I summer is effective against borers. Dipping logs in 50% borax and boric acid for 10 seconds at 55°c, also protects the timber from the attack of *Lyctus sp.* and *Sinoxylon sp.* Seasoning the poles and logs to air dryness are not thereafter attacked by beetles. Debarking the poles also minimise the infection of the wood boring beetles.

B. Diseases

The important disease attacking the Arjun tree are briefly described here. Mortality in seedlings may occur at different stages during their growth and development. The problem due to pre-emergence blight, post-emergence, mortality soon after germination and root rot in older seedlings resulting into poor growth and in extremely cases death may occur. In pre-emergence blight, the developing seedlings are attacked and then killed before the seedlings emerge out of the ground The emergent crop is sparse. This may be wrongly attributed to poor viability of seeds. The fungi spread rapidly in the tissue and the seedling either exhibit wilt symptoms or fall down on the ground due to rotting of tissues at the base. The soil moisture sometimes helps for disease development.

In nursery, leaf spot diseases are responsible to cause major losses due to premature defoliation or leaf blight resulting in to mortality of the plants. Three major leaf' spot diseases are recorded on arjuna. *Colletotrichum arjunae* and *Pestatatiopsis diseminata* causing spots on leaf, *Sphaceloma terminaliae*is responsible for causing leaf and twig blight disease,

Control measures

The damage caused by above leaf spots diseases can be controlled by spraying of 0.2% bavistin at 15 days interval or two spray treatments.

Armillaria mellea causing serious root rot diseases in plantations has been commonly recorded in both young and older plantations of this tree species.

This root rot disease can be controlled by removing the infected plants from the plantation sites. The infected plants can be treated with sodium nitrate, borax and brastical in the root region.

Uses

Arjun is used in many ways. It is a handsome, shade and - ornamental tree. The tree is lopped for fodder in Maharashtra, Madhya Pradesh and Uttar Pradesh and yields fodder of medium quality. They are described here.

1. **Small timber wood:** The heartwood is hard, strong, moderately heavy (specific gravity 0.14, wt. 810-865 kg/cu m), moderately durable, difficult to season, and difficult to work. It is used for carts, agricultural implements, water troughs, boat building and various domestic purposes. It is suitable for plywood (tea-chests) and other structural purposes, such as house construction, water traps, and tool handles.
2. **Bark:** The most important non-wood forest product of arjuna is the bark. The bark is used medicinally and in a number of Ayurvedic preparations by many leading companies like Dabar, Vaidyanath, Maharishi Ayurvedic, Jhandu and Himalayan Drug Company and others. According to an estimate the present, annual production of *T. arjuna* bark is 5,125 quintals in our country, most of which is consumed by the leading pharmaceutical companies and tanneries. On an average, the yield of bark per tree varies from 9 to 45 kg. Its bark has astringent effect and is used in fevers, and in fractures. It is also taken as a cardiac tonic. The bark is also employed for tanning (15.8%) and extensively used in mixture with *Anogeissus latifolia* and *Emblica officinalis*. The fruits also provide tannin ranging from 7-20 per cent.

   The bark is harvested from fully well grown and healthy trees and is collected by chopping the stern bole. The bark is harvested and cut repeatedly to produce fresh crops of bark.

   Once removed, bark grows again well in its original thickness within two years. The yield of harvest can be obtained regularly in 3 years rotation cycle. Harvested bark is sun dried before collection and storage.
3. **Tasar silk worm culture:** Systematic plantations of this species improve rearing conditions for the tasar silkworm through suitable techniques for reducing larval loss and increasing cocoon production. Plantations at 1.2 m apart are made where the foliage yield is about 26 tonnes per hectare. Due

to high plant frequency in tasar bush plantations, supervision for rearing becomes easier, giving a return of about Rs. 4,000 annually on average per hectare. The silkworm rearing on arjun trees is also practiced by tribals in Bihar, Orissa and Madhya Pradesh. The Central Tasar Research Station, Ranchi, has estimated the cost of one hectare of plantation with proper agronomic management yields of about 26 tonnes of leaves annually to sustain the raising of about 32000 cocoons.

GRAINAGE TECHNIQUES IN TASAR EGG PRODUCTION

In adequate production and lack of timely supply of quality commercial tasar seed are major bottle necks in successful utilization of available food plants and optimal production of raw silk. Often non-availability of required seed compels may tribal tasar rearers to skip their crop.

Seed production activities often face scarcity of parental raw material due to wild nature and erratic reproductive behavior of moth. The activity of tasar grainage is complex. Its success depends on the availability of suitable seed cocoons, proper emergence of moths, synchronized coupling of male and female, their disease freeness.

Grainage Technology in tropical tasar silkworm

The main objective of the grainage is to prepare quality disease free layings. Preservation of seed cocoons and grainage operations are two important phases of seed/egg production. A systematic and methodical process of egg production will help in preparation of quality eggs which will ensure successful silkworm crop with optimum cocoon productivity.

Necessity for a grainage or importance of grainage

In a grainage the eggs are produced on scientific lines which is not followed by the rearers.

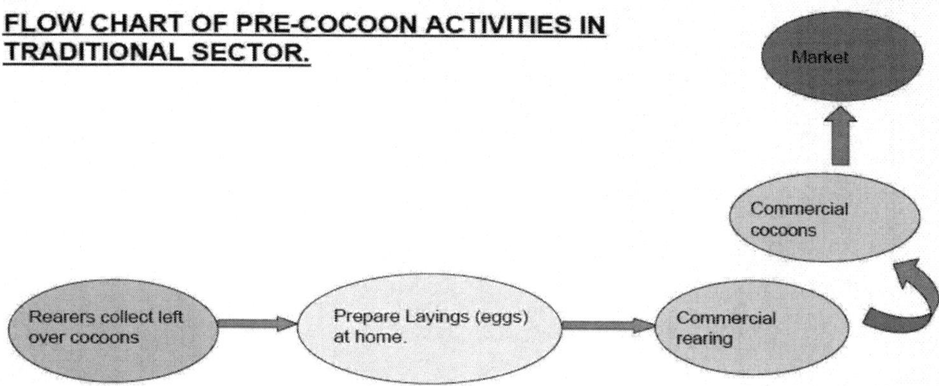

Advantages of grainage

- Ready availability of disease-free eggs.
- Eggs from a grainage have higher hatching percentage
- Timely supply of dfls based on local demand
- Provides supplementary income and self-employment opportunities in the rural areas

Location of grainage

- One should know whether it is traditionally tasar growing area.
- Whether at least **40-50 active** rearers operating in nearby areas
- Whether at least **50%** of the prospective customers are interested in purchasing dfls on cash payment
- At least 5-6 rearers who have the capacity and are willing to carry out seed crop rearing

Size of a grainage unit

- The scale of operations is determined based on several factors like,
  - Availability of cocoons
  - No. of graianages already operating in a cluster,
  - Man-power and resources available with the grainage operator
- Ideally, the scale of graianges should be approximately **5,000 dfls,** which means handling **20,000-25,000 seed cocoons/operational cycle.**

Features of a Grainage house

- The grainage building has special importance, as It is also used to preserve tasar cocoons for next about six to seven months in diapausing stage

— A grainage house functionally has to provide space for: seed cocoon preservation & processing, oviposition room, MME room, egg processing room, egg preservation/incubation room & store room
— Grainage house in an elevated place away from rearing field
— Orient north-south direction, provide a veranda all around the building with rat proof structure
— Fast growing trees all around the grainage building
— Running water facility, to improve cleanliness
— Good power supply to run the grainage activities smoothly
— 12.00 m (l) x 7.00 m (b) x 4.50 m (h) size hall to accommodate upto **1.00 lakh cocoons** at a time **(20,000 – 25,000 dfls** can be produced in one operation
— The grainage room is generally **rectangular**
— The building is constructed on raised ground
— The walls should be at least **60 cm thick**
— Roof should be first thatched, false ceiling should be made
— Facilitate cross ventilation & floor is cemented with mud and lime
— Inner walls of the grainage are white-washed.

Disinfection of grainage house

— Fumigation of Grainage house with two liters of pure commercial formalin by heating
- 5 % formalin is sprayed inside the grainage house
- Well white washed with lime & bleaching powder.

The quality seed production being the prime mandate of grainages, they should adopt available cost-effective knowhow for full utilization of emerged moths. Presevation of seed cocoons, operational procedures and facilities at work place, technological inputs, trained work force and above all, the appropriate scientific handling of emerged parental moths makes commercial tasar seed enterprise successful in attaining optimum productivity and quality.

2. Selection of seed cocoons

All the harvested cocoons are not treated as seed cocoons. It is first selected based on visual and touch. The well built tough and live cocoons irrespective of size are retained in the grainage for egg production discarding, uji infested, yellow fly infested and dead cocoons. This enhances preservation and grainage efficiency.

— Select seed cocoons from selected rearing fields should be harvested after,
  - 4 days-first crop
  - 7 days- second crop
  - 10 days – third crop from the date of onset of cocoon formation
  - High-quality seed cocoons should be selected on the basis of good built and toughness. Should be disease free.

3. Seed cocoon preservation

The selected seed cocoons are **tied in bunch of five, such 20 bunches are made into garland and hung 2' above ground level to a bamboo frame.** The distance between garlands should be 8 to 10" and between rows 1½ to 2' to facilitate grainage operations.

— Sorted good cocoons are tied together with a thread to **make garlands (a garland consists of 100 cocoons)**

— Seed cocoons are hung in the graiange house on bamboo poles with spacing of 15 cm in b/w garlands in the row & 60cm in b/w two rows

— Garlands should be 60cm above the ground level

— A ground area 4.50-6.00 x 3.00-4.00m can accommodate 20,000 cocoons

— Maintain temp. below 35^0 C & RH should not exceed 70 %.

4. Environmental manipulation inside grainage

- High temperature lead to **poor emergence** pr emergence of weak or deformed moth. High humidity leads to **erratic moth emergence.**
- Due to severity of climate, there may be loss of **30-50%** stock, which could be pupal mortality, erratic and unseasonal emergence. Normally winter preservation do not experience any loss due to pupal diapauses but during summer, high temperature, affects the seed cocoon preservation. The optimum temperature should be 35^0C and 40-60% RH to reduce pupal morality and to minimize unseasonal emergence. For which additional cooling measures are to be taken.
  1. The verandah of grainage covered with bamboo mats, hessian cloth curtains or straw pads.
  2. Doors and windows should be covered with wet khus mats. Water is sprinkled as and when required.
  3. Water may be dripped on hessian cloth covered on roof
  4. Coolers designed by CTR & TI may be fixed to windows from outside.

Grainage operations

Grainage efficiency in tropical tasar is not upto desired level. The wild nature of insect poses many difficulties in captivity such as Unsynchronized moth emergence, Low coupling in captivity **(70%), Longer oviposition period (6-7 days), Low egg recovery (80%), Low fecundity (125-175w/laying) and Low Hatchability (60-70%).**

Maintenance of abiotic factors within threshold level will give satisfactory results. Temperature of 20-30^{0}C and > 70% RH and normal day light.

Moth emergence

Usually moths emerge late in the afternoon and continue upto early hours but peak period of emergence is **19 to 21 hours.** Males emerge **earlier** than females. Due to shorter life span and non-feeding stage, the parental koths need much attention while handling to optimize their utility in seed production.

Synchronization of emergence

— Male moths emerged on earlier days are preserved under cold conditions and used for mating

— The moths emerge late in the evening b/w 6 pm and 10 pm

— Nylon net pairing.

Mating of moths

Maintain darkness in grainage room during coupling. Mating takes place after 2-3 hrs of emergence preferably in dark, cool and humid conditions. Mating duration of **4-6** hrs is adequate for ensuring normal fecundity and fertility.

Natural mating occurs inside the grainage, but for increasing efficiency i.e., only 70% female moths undergo natural mating on the first day of emergence remaining 30% go as a waste.

Therefore, nylon net enclosure is proved to be ideal for increasing the mating chances of such females. Male moths can be preserved in refrigerators at **5^{0}C** and can be reused during shortage.

Oviposition

After optimum period of coupling, moths are decoupled and wings of female moths are cut, abdomen is gently pressed to urinate and then allowed to oviposit the eggs. Moths prefer **darkness for** oviposition. **The eggs are ovipositor in batches of 5-10.**

— The moths normally lays about 80% of the total eggs during the first **72 hours** after decoupling

— **Egg laying devices:**

- Earthen cups
- Plastic egg laying boxes
- Nylon net bags

Among all the devices used, **earthen cup is the best for *A. mylitta*. Maximum egg laying is recorded in 16 to 20 hrs. The moths from diapausing crop have a lower fecundity than those from non-diapausing crop.** Eggs oviposited within first 72 hrs. (85-90%) are processed for silkworm rearing.

Mother moth examination.

— By crushing of female moths

— Prick method of moth examination

— The moths are examined for the presence of pebrine spores.

— The eggs laid by only healthy moths are retained.

Disinfection of eggs

After mother moth examination for individual moths, (15% eggs retained in moths), the layings are separated and washed with soap solution and water and then treated with 5% formalin solution for 5 min for surface sterilization. Thereafter, 3-4 times washed with water until formalin smell is removed completely.

Drying of eggs

After proper disinfection of cggs, the eggs are dried under fan in a sterilized room. It is spread on blotting paper and allowed to dry. Care should be taken for perfect drying otherwise there is chance of fungal infection.

Packing of eggs

Properly disinfected eggs are packed in perforated plastic boxes or muslin cloth bags after weighing. Cellular DFLs can be packed in small, perforated paper envelopes of size 3x 2½". The cloth bags of size 9x5" holds **100 DFLs.**

Delaying egg hatching

For delaying egg hatching place them in refrigerator at 8 ± 2^0C for 8-10 days. After withdrawal, the eggs are incubated at $28\text{-}30^0C$.

Transport of eggs

A specially fabricated egg carrying basket may be used for safe transport of eggs to longer distances during summer. It consists of two plastic boxes one inside the other. The inner one holds the perforated plastic jars (50DFLs in each jar), the outer basket harbors the inner box. The inner wall of outer box including cover has a khas mat lining upon which water can be sprinkled during hot and dry weather one such **basket of medium size holds 1800 DFLs.**

The eggs can be retained inside the basket till hatching even after transportation is over.

Seed organization

- Elite seed production at CTR&TI, Ranchi/CTSSS, Lakha, raigarh (Chattisgarh)
- Nucleus seed production at CTSSS, lakha and some selected BSM & TCs involving adopted seed rearers and the use by BSM & TCs (22 nos. located in nine states)
- Basic seed production at BSM & TCs and its supply to state PPCs/NGOs/ seed reares
- Commercial seed production at PPCs/private graineurs and its supply to commercial rearers
- Reelable cocoon production by commercial rearers.

BTSSO, Bilaspur is located at Bilaspur in Chhattisgarh state. The subordinate units, Central Tasar Silkworm Seed Station (CTSSS), Kota and 22 numbers of Basic Seed Multiplication and Training Centres (BSM&TCs) located in nine tasar producing states, are functioning under administrative and technical control of this organization.

TASAR SILKWORMS – BIODIVERSITY – DISTRIBUTION AND ECO RACES OF TEMPERATE AND TROPICAL TASAR

The name Tasar is derived from sanskrit word *trasara* meaning shuttle. It is mentioned in literature dating back to 1590 BC.

Due to the presence of prolific wealth of food plants, abundant man power and ideal climatic conditions, tasar culture offers India a unique opportunity to boost her silk production and economic status of tribal people. About 12 million tribals live in tropical tasar belt and 3 million in temperate tasar belt.

Amongst many wild tasar silkworm eco-races available in the deep forests of Chattisgarh, Raily and Baraf are the most popular varieties for commercial exploitation. Raily cocoons are found naturally in the Sal (*Shorea robusta*) tree forest in southern part of Chhattisgarh. The tribals collect these cocoons from the forest and sell it in the nearest weekly market. Thus becomes a regular source of earning to tribals and they can earn Rs. 2000/. to 3000/. Per annum by collection and sale of nature grown tasar cocoon. Around 44577 tribals are benefited through collection of 500 lakh Nos. of nature cocoons.

1. Biodiversity of tasar fauna

- The important species under the genus *Antheraea* includes the *Antheraea mylitta* - Tropical tasar.
- Close relatives of *Amylitta* are *Antheraea pernyi* - Chinese tasar and *Antheraea polyphemus*.

2. Temperate tasar - *Antheraea pernyi.*

a) The production of this in China dates to Hand and Wei dynasties.

b) Japanese tasar- *Antheraea yamamai*

c) Oak tasar (*Antheraea proylei*)

- The extensive survey conducted in 1966 by the Central Tasar Research Station, Ranchi (Bihar) brought to light the vast wealth of oak in the region. The oak tract extended from Jammu and Kashmir in West to Manipur in the east, comprising Himachal Pradesh, Uttar Pradesh, West Bengal, Sikkim, Assam, Arunachal Pradesh, Megalaya, Mizoram and Nagaland.
- It is a hybrid from uneconomic indigenous species, *Antherae aroylei* and Chinese species *A. pernyi*. It is popular in the north eastern states due to ideal socioeconomic and rearing conditions.

Ecoraces/biotypes/ecotypes/morphovariants of tasar

Extensive surveys conducted in states of Bihar, MP, Orissa, Maharashtra, AP, Assam, Manipur, Meghalaya, Nagaland, Karnataka, Rajasthan, Himachal Pradesh, Jammu and Kashmir, West Bengal, Uttar Pradesh, Union territory Dadar and Nagar Haveli showed the presence of 43 ecoraces. Certain ecoraces are maintained at CTR and T1 Ranchi for breeding programme only six are commercially exploited.

Major economically important ecoraces are distributed in Central India within 18 - 24°N and 80 - 88°E. The ecoraces are restricted mainly in

1. Tropical moist deciduous forest area where average rainfall varies between **1200 - 2000 mm.**
2. Dry tropical forest area - **rainfall is 1000 mm.**

The ecoraces are distributed in five soil types *viz.*, **red loamy, sandy red, black clay laterite and forest hill.**

Tasar occupied varied topographical, climatological, vegetational and edaphically diverse areas and exhibits diversity in phenotypic, physiogenetic and behavioral characters. 3 groups of ecopackets were found in the Bastor forests in Chhattisgarh (Madhya Pradesh).

- **Dabha, Nangoor, Tokapal and Tongpal - Group I**
- **Kondagoan and Narayanpur- Group II**
- **Geedam - Group III**

Significant differences were observed between group I and III. Populations of group II indicated homogenity and heterogenity with the certain populations of ecopockets of group I and III. Thus group I and III are two independant stocks and group II is an intermediate stock of group I and III.

The tropical tasar silkworm has a distribution over wide geographical zones and it has a highly diversified genotype because of their adoption to different ecological conditions and polyphagous nature.

Ecoraces from Bihar

Daba, Sarhihan, Munga, Modia, Laria, Lodhma, Palma, Japla, Kowa, Barharwa. All these ecoraces are found in tropical moist deciduous forest with red loamy soils.

- ***Terminalia arjun (Arjun),* and *T. tomentosa* (Asan)were the predominant food plants for Daba, Sarihan.**
- ***Shorea robusta (Sal)* is the predominant food plant for Munga, Modia, Laria, Lodhma, Palma Kowa, and Barharwa.**
- ***Zizyphus jujaba* (Ber)for Japla ecoraces.**

Orissa

Modal, Nalia, sukinda, Boadh, Simlipal, Omarkote, Sukly

Soil type: Red loam

Forest type: Tropical moist deciduous.

Predominant food plant

-*Shorea robusta for* Kowa, Barharwa, Modal, Nalia, Simlipal, Omarkote, Sukly.

-*T. arjuna* and , *T. tomentosa* : Sukinda, Boadh

Madhya Pradesh

Sukly, Raily, Kurudh, Multai, Mandalla, Jhabura, Bhopal Patnam, Jhabua, Piprai, Seoni, Janghbhir, Korbi.

Soil type: Red loamy except for Raily and Bhopalpatnam for which sandy red is suitable.

West Bengal

1. Tira - *Lagerstroemia parviflora* red loam Tropical moist deciduous forest
2. Bankura- *L. speciosa* red loam Tropical moist deciduous forest
   L. parviflora

Dadar and Nagar Haveli

Dadar Nagar Haveli *T. crenulata* Black clay - Tropical moist deciduous forest.

J & K

Shiwalika	*Zizpyhusjujuba*	Forest hill/alluvial	Mountain sub-tropical

Himachal Pradesh

Palampur	*Zizpyhusjujuba*	Forest hill/alluvial	Mountain sub-tropical

Maharashtra

Bhandara	*T. arjuna,* *T. tomentosa*	Black day	Tropical dry deciduous

Andhrapradesh

Andhra local	*T. arjuna,* *T. tomentosa*	Black soil	Tropical dry deciduous

Uttar Pradesh

Monga	*T. tomentosa*	Red loam	Tropical dry

Assam

Now gong and NE 1, 95. Under red loamy soil of tropical wet ever green forest on *Z. jujuba*

Meghalaya

NE 2 95	*Z. jujuba*	Laterite	Tropical wet ever green

Manipur

Jiribam	*Z. jujuba*	Red laterite	Tropical wet ever green

Nagaland

NG, 94	*Z. jujuba*	Red laterite	Tropical wet ever green

Karnataka

Belgaum	*Hardwickiabinata*	laterite	Tropical wet evergreen

The best performance of these eco races is seen only when they are reared under natural conditions ie, suitable soil, climate and host plants. Higher shell ratio was obtained when reared under natural conditions than under captive conditions.

Characteristics of some important eco races.

Laria

Laria is a wild ecorace of tropical tasar silkworm *Antheraea mylitta* Drury and the adaptability of this ecorace is mainly on *Shorearobusta* (Sal). It has high economic importance and is noteworthy for its **small size cocoon with long peduncle**, robust with variable cocoon colour(Whitish grey, grey, dark & light yellow) with **low filament denier.** In nature, the voltinism of this ecorace is uni, bi-, and trivoltine.

In some limited areas, farmers use Laria ecorace for 2nd crop (during September-October) that too in small scale. Adaptability of this ecorace is primarily on *Shorea robusta* (Sal) and this flora covers 86.9% of the total tasar flora in India as a gainst 13.1% of *Terminaliaarjuna* (Arjun) and T. tomentosa (Asan).

Modal

Modal **ia a univoltine race found in Similipal hills of Orissa**. It spins the cocoons by the care of nature and complete the first life cycle. The race is characterized by **highest fecundity of eggs, unique moutinism, varying colours of cocoons (Greyish black, Dirty white, cream white, pinkish grey**

and light yellow) with strong and stout peduncle having one **ring,** thick shell with higher shell weight and higher silk content and totally disease freeness.

Generally, a greater portion of these cocoons are collected from the wild habitat by the local tribals and the cattle herders and sold to the Tasar reares cooperative Societies. (TRCS). The TRCS finally sells the cocoons to the interested rearers.

The left-over Modal races in the Jungle remains from August to next year May –June for repeating the life cycle again. At reares level, the dfls are prepared, rearing is conducted and the cocoons are harvested.

These cocoons are known as **Bogai** and are characterized by thick and small cocoons with comparatively less cocoon weight, shell weight and silk content than Modal. Likewise, a second crop od Bogai is also possible but the cocoon characteristics deteriorate. The modal eco race is univoltine in nature, but when the rearing is conduced outside its, natural environment, it behaves as trivoltine.

Nalia

It is also found in **jungles**. The race is characterized by **slender and long peduncles with two to three rings.** In all other aspects it is **similar to modal.** First life cycle is completed in jungle and the cocoons thus spinned are collected by the local tribes and cattle herders.

Daba and Sukinda

These races are both bi voltine and tri voltine. **In case of bi voltine,after completion of first life cycle, the pupa undergoes dia pause and in case of trivoltines, the pupa undergoes diapause after third life cycle.** In the nextyyear, when the monsoon sets in, the life cycle starts for both bi and tri voltines.

These races are commercially exploited and three rearingsin case of tri voltine and two in the case of bi voltines are conducted annually.

The first crop rearing starts by last week of June and takes 33 days. It is also known as Ampatia crop. The second crop rearing generally starts by September and takes about 40 days. The first and second rearings usually suffer a great loss due to cyclonic weathers with storms and heavy rains. high temperature and humidity causing diseases and different kinds of predators. **The first and second crop are known as seed crops as the cocoons harvested are**

utilized for seed production. The third crop rearing is conducted for trivoltine race called Jadai crop. The rearing starts by fourth week of November and takes 60 -70 days. The cocoons are harvested during January. These cocoons under diapause upto May.

Performance of different eco races of tasar for shell ratio under naturally grown and reared condition.

Ecoraces	Naturally grown		Reared	
	Male	Female	Male	Female
Daba	16.76	14.20	12.89	11.37
Sukinda	16.00	14.77	12.63	10.55
Raily	20.35	16.52	11.41	10.17
Modal	24.51	19.90	11.73	10.66
Laria	21.11	20.81	12.07	9.74
Sarihan	14.50	13.73	11.79	10.02
Bhandara	20.09	19.43	11.05	10.64
Andhra local	16.02	17.85	13.38	11.08
Tira	16.35	15.35	17.81	14.91

The eco races tested for semi domestication status were Daba, Sukinda, Modal, Raily, Laria, Andhra local, Bhandara and Sarihan.

The best performance of an eco-race is seen only when they are reared under natural conditions *ie*, suitable soil, climate and host plants. Higher shell ratio was obtained when reared under natural conditions than captive conditions.

Semi domestication status of commercially exploited ecoraces

The races tested for semi domestication status were **Daba, Sukinda, modal, raily, Laria, Andhra local, Bhandara and Sarihan,**

Ecoraces	Semi domestication status	Observation
Daba, Sukinda Modal,	Easily amenable	Wide range of habitat preference
Raily, Laria	Least amenable	Silk richness maximum under natural Sal forest and quality deteriorates in capture condition
Andhra local, Bhandara and Sarihan	Poor adaptability	Poor commercial characters

Among the eco races tested, **Daba and Sukinda** races are alone found to be suited for commercial exploitation.

Inbred lines evolved at CTR &TI (Central Tasar Research and Training Institute)

For maintaining purity of materials, a no of inbred lines was evolved at CTR & TI, Ranchi. GB 2, HB 3, GB 4, 5, 6,7, 8, 9, GE 1, GE 2, GF 2, GF 3, GF 5, GB 913, GB 4 14, GB 4 15, GBB 9 16, GB 5 11, GE 212, R 57, L8, RS, S 17, RF 1, RF 4, RF 35, Nagri 1, Nagri 2.

Commercialization of three tasar ecoraces

- Daba TV,
- Daba BV,
- Sukinda BV

BIOLOGY OF TASAR SILKWORM

Tasar

China is the largest producer of tasar followed by India. Tasar producing states in India are Jharkand, Chattisgarh contributing 70% of the production, Andhra Pradesh, Orissa, West Bengal in appreciable quantities, while U.P, Maharastra, Bihar and Madhya Pradesh are minor producing states.

Food plants Primary: Terminalia tomentosa, Terminalia arjuna.

Secondary: Terminalia catappa, Zizyphus jujube.

Egg

The egg is oval, dorsoventrally symmetrical along the anterio posterior axis, about 3mm in length and 2.5 mm in diameter. It weighs approximately 10mg at Ovi-position and it is dark brown in color. Two brownish parallel lines along the equatorial plane of the egg divide the surface into three zones, disc, streak and edge. Eggs undergo incubation for 3-5 days.

Larva

The larva is typically cruciform and has a hypognathous head with biting and chewing mouth parts. On hatching it is dull brownish yellow with black head. The body normally turns green and the head brown after 48 hours, but also yellow, blue and almond colored larvae are seen occasionally. The size and weight at maturity are about 13 × 2.1 cm and 50gms.The larva passes 5 instars within a period of 26-28 days for I crop, 42-45 days II crop, 55-60 days III crop.

The prothoracic hood of the first instar larvae dorsally bears an oval black spot, which early in the second instar becomes M-shaped and then, later on V-shaped with two dots. These marks are absent in the third instar, but reappear in the fourth and fifth instars as two semi lunar red markings.

Third-instar larvae have a yellow lateral line extending from the second to the tenth abdominal segment. This line is bordered by a brown upper line in fourth and fifth-instar larvae.

Tubercles: The body of silkworm larvae has prominent protrusions or out growths on thoracic and abdominal segments. These out growths are called tubercles. They work as sensors for changes in environment related to temperature and humidity. Based on the position five types of tubercles are recognized. Dorsal (DT), upper lateral (LT), lower lateral (LLT) and caudal (CT). They are black in the first instar, orange red in the second and violet in the third to the fifth instar.

Hairs and Setae: The larval body is covered by hairs and setae. The hairs are white, minute and irregularly distributed over the body. The number and arrangement of setae vary according to the type of tubercle, body segment and age of the larvae. Generally, the total number of tubercular setae remains constant until the third instar and diminishes thereafter.

Shinning Spots: Silvery white lateral shinning spots, either oval or triangular, appear during the third instar at the foot of upper lateral tubercles of the second to the seventh abdominal segment. Six regular and thirty-eight irregular patterns have recorded. Plain larvae are also common. Other shinning spots are present at the base of the dorsal tubercles. These are brick red in green and yellow larvae and white in blue and almond-colored larvae

- First brood (July-Aug) : 30-35 days
- Second brood (Sep- Oct) : 40-45 days
- Third brood (Nov- Jan): 60-70 days

Stage	Duration (days)
Egg	7-8
Larva	29-43
Pupa	30-35
Adult	8-10
Total	**74-96**

Morphological features of Tasar silkworm egg

Features	Tropical tasar	Oak tasar
Colour	Brown, light yellow or creamy white	Bluish green
Shape	Ovoid	Ovoid
Symmetry	Bilateral	Bilateral
Size (mm)	3.0 x 2.8	2.8 x 2.2
Weight (mg)	10	8
Lines/streaks on egg surface	Two brownish parallel lines	Absent

The surface of egg is divided into three zones viz.,

- Disc zone,
- Streak zone and
- Edge zone

Colour of larva

Silk worm	On hatching		Later stages	
	Head	**Body**	**Head**	**Body**
Tropical tasar	Black	Dull brownish yellow	Brown	Green, yellow, blue, almond
Oak tasar	Red	Black	Dull brown with ten dark brown spots on head capsule	green

Size of larva

Larvae	On hatching		At maturity	
	Weight (mg)	**Size**	**Weight (mg)**	**Size**
Tropical tasar	8.0	7.0 x 1.0	40 ± 5	13.0 x 2.1
Oak tasar	6.5	6.5 x 1.0	15 ± 5	8.5 x 2.0

Pupa and cocoon

- In Tropical tasar, the cocoon is single shelled, oval, closed and reelable, having a hard non-flossy shell with fine grains
- The cocoons are generally yellow or grey
- The dark brown pupa measures about 4.5 x 2.3 cm and weighs 10 ± 2 g
- In case of Oak Tasar, the cocoon is single shelled, oval, hard, non-fibrous and reelable
- The pupa weighs about 5.46 g and measures 3.5cm x 1.80cm.

Tropical tasar *(Antheraea mylitta)*

- The females are bigger (4.5cm), with a distended abdomen and narrow bipectinate antennae (1.5cm long)
- The males are smaller (4.0 cm) with narrow abdomen and broad antennae
- The females are polymorphic in colour, being grey and yellow, whereas males are brown.

Oak tasar *(Antheraea proylei)*

- In case of oak tasar, the body length of moth in males 3.5 cm and in females, it is 4.0 cm.
- The wing span in males and females is 14.0 and 16.0cms, respectively.

VOLTINISM OF TASAR SILKWORM

Tropical Tasar *(A. mylitta)*

Life cycle

The life cycle of all the varieties of tasar begins with the onset of rainy season (July - Aug). In *univoltine* race, the life cycle is repeated only once in a year during rainy seasons (July - Aug).

Bivoltine race is cultivated during *rainy period* (July - Aug) and autumn (Sept - Oct.).

Trivoltine repeated during

Rainy (July - Aug)

Autumn (Sept - Oct)

Winter (Nov. - Dec.)

In univoltine diapause extends from last part of Aug to last Part of May

Bivoltine → Nov - end of May

Trivoltine → Jan - end of May

Voltinism of tasar moth in different altitudes

Altitude plays a significant role in the voltinism of tasar silkworm. At low altitude, voltinism is high and it is low at higher altitudes. *A. mylitta* is trivoltine at low altitude and bivoltine at medium altitudes. *A. mylitta* is trivoltine at low altitude and bivoltine at medium altitude. *A. mylitta* is not found at higher altitude. *Antheraea paphia* is bivoltine / trivoltine at low altitude (50 - 300 ASL) bivoltine at medium altitude (300 - 600 m ASL) and univoltine at high altitude (601-1000 m ASL).

Altitude (MSL)	Local name	Voltinism	State of cultivation	Frequency of life cycle
50 – 300	Bogei	TV or	Cultivated	Three
	Sukinda	BVTV	Semi demonstrated	
301-600	NaliaDaba	BVBV	Wild semi domesticated	Two
601 – 1000	Modal	UV	Wild	One

Emergence of tasar moths

Tasar moths remain under pupation for minimum period of six months ie from January to June during extreme coldness (late winter January) and extreme hotness in summer (April and May intercepted by spring (February and March).

Generally, all the resting pupae emerge during rainy period (June ie on the onset of monsoon. The moths from first crop harvested in July emerge during August - September.

Trivoltine

Second crop harvest commences during October followed by further emergence during the last part of October and early November for third crop.

Temperature and humidity decides the moth emergence. 26 - 28°C temperature and 72-80% RH aid in regular emergence of moths.

Eggs are oval, dorso ventrally flattened and bilaterally symmentraical along the enter posterior axis. Each measures 3mm in length and 2.5mm in diameter, weighing 10mg. Eggs are white, light yellow or creamy. Two brownish parallel line along the equatorial plane of the egg divide the surface into three zones (disk, streak, edge).

Egg period: 3-5 days

Larvae are eruciform and possess a hypognatus head with biting and chewing mouth parts. Newly hatched larva is dull brownish yellow with black head, measures 7 x 1 mm and weighs about 8 mg. after 48 hrs. Larval body turns green and head becomes brown. Occasionally yellow, blue and almond coloured larvae are also found. Mature larva weighs 50 grams and measures 13 x 2.1 cm. Each larvae passes five instars.

The larval duration

- 26-28 days (first crop);
- 42-45 days (second crop);
- 55-60 days (third crop).

The prothoracic hood of the first instar larvae dorsally bears an oval black spot, which early in the seconds instar becomes M-shaped, later on V-shaped with two dots. These are absent in third instar, but appear in the fourth and fifth instars as two semilunar red markings. The anal flap has a triangular black mark early in the first instar, which becomes V-shaped and brownish from second instar onwards.

Sexual marking appears lateral in the fifth instars as milky white spots on the ventral surface of the eighth and ninth abdominal segments. In all instars the larvae possess five types of tubercles (dorsal, upper lateral, lateral, lower lateral, caudal).

They are black in the first instar, orange red in the second and violet in the third to the fifth. White, minute hairs are d distributed irregularly all over the body. Setae are of two kinds. Silver shining spots appear during the third instar of second to seventh abdominal segments.

Pupa is obtect adectious with segmented, dark brown colour body measuring 4.5x2.3 cm and weighs 10.3 grams. The sexual markings are on eighth and ninth segments.

Cocoon is single shelled, pendent, oval, closed and reelable, non-flossy with fine grains the anterior end has dark brown peduncle with a ring at the distal end. Cocoons are yellow or grey.

Pupal period

- In univoltine diapause extends from last part of Aug to last Part of May
- Bivoltine → Nov - end of May
- Trivoltine → Jan - end of May

Adults

Moths exhibit sexual dimorphism. The females are bigger (4.5cm) with a bipectinate antennae (1.5cm long) and broad abdomen. Males are smaller (4.0cm) with broad antennae and narrow abdomen

Male wing span is about 16 cm. the area of fore wing is about 2121 mm2 with a centrally positioned ocellus (70mm2). While hind wing is about 1584 mm2 with an ocellus of 50mm2.

Female wing span is 18 cm. The fore wing is about 2350 mm2 with ocellus (85mm2). The hind wing is 1850mm2 size with ocellus (25mm2) The colour of ocelli is same in both sexes. Wing scales, wing venation are specific.

CHAWKI REARING METHODS OF TASAR SILKWORM

Tasar culture is a subsidiary occupation of tribal inhabiting in and around the forests of Bihar, Madhya Pradesh, Orissa, West Bengal, Andhra Pradesh and Maharastra.

The tribal were associated with collection of cocoons initially and later with cocoon production through planned rearing in forest patches.

The eggs can also be obtained from the private grainages.

The process of rearing includes incubation of eggs, brushing of neonate larvae, feeding, safeguarding against diseases, pests and predators and care during moulting.

Rearing of *Antheraea mylitta.*

Incubation of eggs

For proper embryonic development, **70-80% RH and 28-30^{0}C** temperature is optimum. It is kept in wooden or paper tray.

Brushing of neonate larvae

- It is the transfer of larvae from egg box/ egg tray/bag/cup to the fresh succulent leaves of primary host plants.
- **Normally eggs hatch in the morning hours.**
- The leaves are spread on the tray and neonate larvae crawl on to them in one to two hours.

Feeding young larvae

After brushing, young larvae are transferred to chawki garden under nylon net cover. The rearing capacity of well-developed economic plantation of 4 years and above is 450 DFLs (140 plants for Asan) which may sustain rearing upto second instar. (Nylon net size of 40'x30'x10' can cover 70 plants). **Plants are pruned at 3' height during January-February**.

Temperature requirement

Stage	Temperature(°C)	Humidity (%)
I instar	26-28	80-90
II instar	26-28	84-90

Important activities to be done at the rearing site

- To avoid harbouring of harmful insects like ants, pests and predators, the chawki garden floor and surrounding should be kept weed free.
- Bleaching powder @ **10g/ plant should** be sprinkled on floor once or twice.
- **2 per cent formalin** should be kept at the site on the stand in the earthern/ plastic basin for disinfection of hands.
- Regular removal of litter from floor is essential by sweeping. The area/ floor is also covered with polythene sheet. Later the sheet containing the excreta of the worms are removed carefully.
- Dead and discarded worms should also be removed from the site and burnt along with apical leaves.
- All the rearing appliances are disinfected with **5% formalin spray.**
- For free air circulation, nylon net may be raised frequently an all sides and kept as such for 15-30min. After rain, the net should be shaken for removal of water film which restricts the free air flow.

Sometimes due to unsuitable climatic conditions like continuous rain, stormy wind, dry and hot temperature, it is not preferable to rear under nylon net. In such conditions, indoor rearing may be resorted.

The most suitable indoor rearing methods are

1. Bottle rearing
2. Pitcher rearing
3. Tray rearing
4. Sieve rearing

Bottle rearing

In this technique, newly hatched larvae are brushed in a tray on the leaves of cut twigs of Asan (*Terminalia tomentosa*) or Arjun (*Terminalia arjuna*) or Sal (*Shorea robusta*). After brushing, the cut ends of three to five twigs of about 2 feet length are inserted in a bottle containing water.

The twigs are then covered with a polythene bag supported by a bamboo frame. The leaves of inserted branches are folded and clipped to avoid touching the polythene enclosure and prevent worms crawling onto it.

One end of the polythene is tied round the neck of the bottle and the other to a support. On each such set, **worms hatched from 1-5 dfls can be reared**. Invariably, single day hatched worms should be mounted over a set to ensure better handling of the larvae for uniform moulting. A hut of 10' x20' size can accommodate 90 such sets in two tiers (15' x 3' x 2') covering about 300-400 dfls.

The larvae brushed on bottle sets are allowed to feed on the leaves till they settle for first moult without effecting any transfer. The lower end of the polythene enclosure is opened once or twice daily for about 15 minutes for aeration and litter cleaning.

The polythene cover provides the larvae, protection from pests, predators and adverse climatic conditions prevailing outdoors and also helps to keep the leaves fresh and succulent for longer duration.

The polythene cover should be removed as soon as the worms start settling for moult, for smooth ecdysis. After the first moult, the worms along with twigs are transferred to outdoor host plants in cool hours (early morning or evening). The worms then crawl onto the fresh leaves of the plant.

Advantages

- This indoor rearing technique does not demand elaborate supervision as the larvae are protected from pests, predators and adverse climate by the thatch and polythene enclosures.
- It is easy to maintain required temperature, relative humidity, leaf quality and hygiene.
- The larvae require only **one feed during the 1st instar.**
- Appropriate quality of leaf provided to the larvae during 1st instar ensures better health and survival in later instars.

- Loss of larvae during first **instar is reduced from 30% to 5-10%.**
- The technique helps to improve **survival of worms** and cocoon yield.
- The technique is suitable for **rainy season**, when direct outdoor brushing causes loss of larvae due to heavy showers.

Limitations

- The technique is suited well for **small rearings** only. For large scale rearing, more thatches and rearing sets are required which involves considerable investment.
- It requires skilled manpower.
- If the weather does not improve within **3-4 days**, further maintenance of this type of rearing becomes cumbersome

Pitcher rearing

Specific structure of **earthen pot commonly known as pitcher** is used for **young age** rearing.

- Small pitchers @ one pitcher per **10 DFLs** are used
- Then **3'ft long twigs** of Asan or Arjun or sal are cut and inserted in the mouth of pitcher filled water up to neck, the cut ends are dipped in water for about 9". **Eight to ten twigs** per pitcher may accommodate the desired number of larvae. Left over space in pitcher is covered with straw to check drowning of worms
- In a very dry spell, the pitchers may be kept and covered with suitably stitched polythene enclosure
- This set up is continued upto second stage.

Advantages:

- Operation becomes **easier** than bottle rearing.
- It has **higher rearing capacity** as compared to bottle rearing.
- For proper **aeration,** no extra efforts are required.
- A thatch of **10' x 20'** size can accommodate about **40 pitchers** covering **400-500 dfls.**

Limitations:

- The tiny larvae may fall on the straw and crawl inside or outside the pitcher. This may lead **to starvation and death.**
- Large scale transfer of worms to outdoor is cumbersome.

Pit Rearing

This type of rearing is done below the **ground surface in pits of size 3' x 2' x 1'.** In this technique, newly hatched silkworms are brushed on the twigs and placed in the pits. The pits are covered with polythene sheets.

The pits are protected with a thatch from rain, wind, storm etc. It is advisable to provide two feeds of Asan/Arjun/Sal shoots during morning and evening, every day. Litter is cleaned early morning and the larvae are exposed to free air for about 20-30 minutes.

Abut 150-200 dfls can be reared in 6 pits of 3' x 2' x 1' dimension upto 3rd instar. After the 3rd moult, the worms are transferred to regular outdoor plantation.

Advantages:

- Rearing arrangement in the pits checks the **temperature fluctuation** and ensure suitable conditions for young age silkworm.
- The polythene cover helps to keep the leaves fresh and succulent.
- The technique helps to **minimize the mortality loss of worms and improves cocoon production.**
- It provides **sufficient space** for movement of the worms.
- Handling is easy.
- Pit rearing is good **during hot summer specially** to rear dfls prepared out of unseasonal emergence when outdoor climate is not congenial for rearing.

Limitations:

- It is an outdoor rearing technique and therefore, enough **vigilance** is required. Otherwise, snakes, lizards and other pests may eat away the worms.
- During rainy days, water **logging** may cause problem in the pits.
- It is not suitable for **commercial/large scale** rearing as number of pits are required to be dug.
- This technique requires watch **and ward from dawn to dusk**

Sieve Rearing

In this technique, a circular sieve (diameter -1') is used for silkworm rearing. The newly hatched silkworm are brushed on tender leaves/shoots of Arjun. The rearing in sieve is conducted upto 2nd instar in a room.

The larvae fed two times a day i.e., morning and evening. One such sieve can accommodate about **500-600 worms (4 dfls)** upto 2nd stage. After 2nd moult, the worms are transferred on regular outdoor food plants (*T. tomentosa* and *T. arjuna*).

Tray rearing

In this technique, wooden trays of **3' x 2' x 4"** size with iron net at the base are used for indoor rearing. Each such tray is provided with cris-cross wooden/ bamboo bars above the base for placing shoots on it. The trays are arranged in shelves of stand made up of bamboo/wood/iron.

Single day hatched worms are brushed on Arjun or Asan shoots with leaves in trays and the worms are fed tender leaves with shoot twice a day i.e., morning and evening. Bed cleaning is done once a day.

Wet foam pad strips and paraffin papers/polythene sheets may be used at the bottom and top to maintain microclimate and keep the leaves fresh and succulent. After 2nd moult, the worms are transferred outdoors on regular host plants during cool hours.

Advantages:

- This indoor rearing technique **saves space and manpower.**
- Required **temperature and RH can be maintained** easily in the rearing room.
- It helps to protect the worms from **parasites, predators and vagaries of nature.**
- Desired quality of leaves can be fed to the silkworms.
- Close supervision with minimum manpower is possible.
- It does **not require watch and ward from dawn to dusk.**
- It improves **survivability and cocoon yield.**
- This technique has been found better than other techniques.

Limitations:

- Tasar silkworm being wild in nature does not stick to the trays and start moving outsides the tray.
- Indoor rearing requires lot of rearing appliances.

Chawki rearing on chawki garden under nylon net (outdoor) :

Chawki garden in tasar culture is a new concept.

Chawki garden with **four years** and above *Terminalia arjuna* plant (4' x 4') are pruned at 3' height and are given input @ 4 kg FYM/plant/yr and N:P:K in the ratio of 100:50:50 kg/ha/yr in two split doses.

The newly hatched larvae brushed on these bushes and thereafter, the bushes are covered with a nylon net (40' x 30' x 10'). Such a net can accommodate 70 chawki plants which can sustain **225 dfls** upto 2nd instar.

After 2nd moult, the larvae are transferred by cutting branches from these bushes to the regular economic plantation. Since this is the most effective way of rearing, a separate package has been developed.

Advantages over Alternative Technologies:

- The nylon net cover of the chawki garden provides protection to worms from parasites and predators.
- Pruning at 3' height is convenient for young age rearing and yields maximum foliage **(4-6 kg/plant).**
- Application of FYM and fertilizer improves the quality of tender leaves.
- Larvae fed with nutritious leaves in young age develop better resistance against diseases in late age.
- This technique helps to minimize mortality/loss of larvae during young age and improves cocoon production by **10-15 cocoons** over all other methods discussed above.
- This technique helps to reduce manpower requirement.
- It also helps to improve **qualitative characters of cocoon i**.e., cocoon shell and pupal weight. Since **fecundity is positively correlated with pupal weight,** fecundity automatically gets improved. Thus, this technique can well be used/adopted by **BSM&TC**s/seed production centres/state govt, farms.
- This technique can help to increase production by 10-15 cocoons/dfls over rearing on economic plantation without net and other inputs.

Limitations:

- Initial high cost which a common rearers may not afford.
- During rainy season, after rains, a water film is formed on the nylon net which may result in **stuffy atmosphere inside** the net that may lead to loss of worms due to **virosis.** Therefore, after rains, the net should be shaken well to remove water film and allow free air movement.

The bamboo poles may be erected properly to avoid uprooting during storm.

LATE AGE REARING METHODS OF TASAR SILKWORM

Late age rearing

Late age rearing starts after second moult. The process starts with the transfer of chawki worms to economic plantation or to forest plants as the case may be.

The transportation of worms is done with specially designed tripod stand. For these three wooden sticks of about 5' length are tied below one feet of upper end so as to form tripod stand.

The upper open ends of these sticks are then tied with a stick ring of bamboo or any other bendable material available in forests to serve as hanger for worms on cut twigs. After hanging the cut twigs, the worms are carried to the place of transfer.

For commercial crop, both economic and forest plants are sprayed with 1.5% urea, 15 days before utilization of plants. One litre solution is sufficient for 5 plants.

Crawling down of worms is common feature in tasar, for this plant trunk is wrapped with polythene strip smeared with grease on outer surface prevents the worms crossing the barrier.

Likewise to check ant attack, grease and BHC 50 WP mixture / methyl parathion in a proportion of (15:15) is used as smear.

It is effective for seven days even in torrential rain . 12 to 18" above ground level is suitable for use of strip. Later, transfer of worms is done from one bush to other without waiting for total consumption of leaves so that 20-25% of leaves are left to maintain the plant requirement and ultimately for better maintenance and stability in plants.

The unwanted heavily branched shrubs available in the forest area are used for collecting the worms to be transferred. After mounting sufficient number of worms, the branch stick is hanged on new plant and worms gradually crawl on to leaves leaving branched stick for further use.

Rearing house should be such where optimum environmental conditions required for silkworm.

Stage	Temperature(°C)	Humidity (%)
III	26	80
IV	24-25	75
V	23-24	70

Cocoon formation

- After voracious eating in the 5th instar, the mature silkworm forms the cocoons on the branches of their host plant.
- The spinning of cocoon takes 4-6 days. The larva inside cocoon transform to pupa in another 4-6 days.
- The cocoons are fit for harvest after pupa is formed.
- Sometimes cocoons are harvested by snatching cocoons (Raily race) which may damage the cocoon as well as ring and peduncle is wasted (which can be used in spun yarn- Balkal). Therefore, care should be taken to cut the cocoons along with twig near the ring.

Mounting and spinning

- As sufficient foliage is required for hammock formation during spinning, it is advised to keep a chunk of plantation separately for spinning.
- One to days before spinning, the worms are transferred to these plants.
- Spinning worms should not be transferred as it leads to non-formation of cocoons.

Harvesting and storage of cocoons

The cocoons should be harvested after six days of spinning. During winter, it is still delayed. Harvesting includes cutting of twigs with sharp knife, separating the cocoons from the twigs, removal of leaves and sorting of cocoons. Flimsy, uzi and ichneumonid infested cocoons are separated from good cocoons. Immediately after harvesting, cocoons are sun dried for some time garlanded and hung in a grainage house if for seed rearing and sun dried it for commercial use.

Difference between conventional rearing and modified rearing

By the conventional method of outdoor rearing, the crop loss takes place to the tune of 50-55% due to pests and natural calamities and 30-35% due to disease.

Therefore, a rearer hardly harvests a crop of 10-20%.

The bulk of these losses may be attributed to the faulty selection of rearing spot and feed plants, improper handling of larvae, poor supervision of the rearings and above to the superstitious beliefs of the tribals.

In 1964, the outdoor rearing was for the first time viewed from a scientific angle and essential prerequisite for success of a rearing were analyzed which are described below :

Selection of rearing site: For a healthy rearing, it is essential to select a spot, which is not low lying. A shady place is as harmful as an excessively humid spot.

Selection of plantation: The fate of rearing largely depends on the type of plantation used. The rearing spot as such must have a good concentration of medium sized (10-12 ft) trees. Pollarding of trees at the height of 6-7 ft results in luxuriant flush of quality leaves.

Preparation of rearing: Quite a lot of preparation is necessary before start of rearing. Rearing spot is cleared of weeds, otherwise would invite pests. The selected plants are freed from insects, unwanted leaves (yellow, dry etc) and the ground below is cleaned. The branches and small twigs on the trunk upto the height of 4’ from the ground should be removed. A band of straw should be wrapped round the trunk of each bush to act as a barrier for the larvae. The rearing appliances viz., bamboo baskets, mounting brushes, spades, secature, basins etc., should be made available before brushing.

Brushing: In the conventional method, leaf cups containing eggs to the host plant branches are used. The larvae on hatching, crawl on to the leaves by their natural instinct. In this method rain water, fluctuating temperature and humidity conditions affect adversely the developing embryo and reflects on hatchability and the subsequent rearing performances.

In the modified system, small twigs of the food plant bearing soft and tender leaves are lightly put over the hatched larvae in the egg boxes to which they quickly migrate. The twigs are then tied with the branches of the food plants at different places to ensure uniform distribution of larvae.

The larvae hatched up to 48 hr only are considered for rearing to ensure a uniform growth and healthy crop. The time suitable for mounting is early morning and evening, or any time in a shady place.

Mounting should be avoided during heavy shower, gusty wind or storm. Each plant should be numbered and in turn should bear a wax coated label indicating the lot number and its origin.

Maintenance of larvae

Supervision: After mounting, the larvae are left undisturbed. Constant watch from dawn to dusk is indispensable to avoid the pest attack. Different methods are applied to scare away the birds according to one's convenience.

Cleaning: The dead larvae are collected twice a day, in the morning and evening hours. They should not be allowed to remain hanging on the branches. The material thus collected are buried in deep pits away from the rearing spot. Slackness in cleaning operations, may cause disease outbreak – a reason for which the tribal's rearing, suffers.

Microscopic examination: Regular microscopic examination of the dead and diseased larvae is essential to detect the diseases.

Handling of infected lots: The close observations of the larvae on each plant are made every now and then to assess their condition. It is desirable to rear the weak and unhealthy larvae separately on fresh bushes. The same is applicable to the lots revealing high incidence of viral or bacterial mortality. Special attention is paid if microsporidian infection is detected in a lot.

Handling of moulting larvae: The larvae under moult are handled with utmost care. Disturbance or transfer of the moulting larvae may cause the rupture of skin or loosening of grip. The transfer should be effected after 2-3 hr of ecdysis when they resume their normal activities and the integument assumes normal consistency.

Quality of leaf: Unfortunately, in outdoor rearing, very little can be done to provide the quality of leaf suited to different instar larvae. All the same, the leaves of the spring sprouting suit better to the requirements of the larvae during the first crop. For the second and third crops however, leaf plucking or light pruning is desirable.

Transfer of larvae: When foliage on a plant is consumed, the larvae are shifted to another plant. The rearers do it by picking the larvae individually and putting them on cut branches of the food plant, which are subsequently attached to a fresh tree. Besides contaminating the whole lot, it inflicts injury to the individual larva while detaching it forcibly. Cutting of small branches carrying the larvae with the help of secature is desirable.

Handling of spinning larvae: For the spinning larvae it is necessary that the plant has fair amount of foliage. The latter, besides lending support for hammock formation acts as camouflage against predators. The larvae under the first phase of cocoon formation should not be disturbed or touched by hand lest they desert the hammock.

Density of larval population: The tribals mount heavy population of larvae on a tree irrespective of the quantity of foliage. This results in overcrowding and frequent transfer of the larvae, causing thereby improper nourishment, stunted growth and heavy disease and mortality. It has been shown experimentally that with the increase in the density of larval population the effective rate of rearing, cocoon weight, shell weight and shell ratio come down, while the disease mortality goes up. Therefore, optimum number of larvae are to be brushed on a tree which can complete cocooning without any transfer should be visualized before hand.

Harvesting: The cocoons are harvested on the sixth day of the initiation of spinning. It is carried out by cutting the portion of the twig near the cocooning. The normal and inferior cocoons should be harvested separately. After harvest, cocoons are cleared of the adhering leaves. While good cocoons are sent to the grainage along with rearing history of the lot, the poor cocoons are stifled and sent for reeling or spinning.

Disinfection: All the rearing appliances are disinfected 24-28 hr before the start of rearing, and after every use they are thoroughly washed with water followed by disinfection. Similarly, every worker in the field must disinfect their hands immediately before and after handling the larvae or cocoons. For all disinfection purposes 2-5% formalin solution is used.

Rearing of *Antheraea proylei*

Generally, two types of rearing are done in *A. proylei*

Complete indoor rearing is practiced in North West and outdoor rearing is in North East.

Out door rearing of *A. proylei* in North East

In the outdoor rearing under natural conditions, the larvae after hatching are mounted on small twigs which are placed in tin/pitchers/bottles and after the first moult transferred to the oak bushes outdoors. They are allowed to grow there under proper watch and ward to check the attack of pests and predators.

Proper hygienic conditions are maintained at the rearing sites. Use of nylon nets ensure better effective rate of rearing. Moreover, it checks the attack of pests and predators to some extent and saves them from natural hazards.

Indoor rearing of *A. proylei* in North West

Since the leaf quality plays significant role in successful rearing, it is necessary to cor relate the rearing with natural leaf sprouting. About 1.4 lakhs of oak tasar food plants are available in Himachal Pradesh.

The oak tasar flora is mostly confined to the hills and is under forest Department. Out of several species of *Quercus,* the abundantly available ones are *Q.incana* and *Q. semi carpifolia* . Oak tasar is distributed at 1000 -3500 m above mean sea level. Quercus flora is mainly distributed in forest areas of Kangra, Chamba, kinnaur, Kullu, Shimla and Lahual and Spiti districts.

Q.incana leaves matured quickly after sprouting within 20 -45 days. *Q.incana* sprouts with maximum leaf production during first week of March at low altitudes. The sprouting time increase with the increase in altitude. Accordingly, the rearing period is in the month of May –June and hence the approximate time for brushing is Last week of May to Second week of June. The rearing period can be fixed based on the sprouting of leaves.

Complete indoor rearing from first instar to spinning is practiced because of lofty oak trees and favourable environmental factors.

The eggs should be incubated at 24°C to 26°C temperature and 80-90 % humidity in dark condition up to four days and the rest 6-8 days under normal room light. Incubation period is 8-10 days. Generally, the larvae start hatching in the morning and continue for 4-5 hrs.

The newly hatched worms are allowed to crawl on small twigs which are placed on the ground itself with waste paper beneath or in the wooden trays with waste paper beneath or onto the cut twigs or branches inserted in the bottle / tins/ pitchers as the case may be. In the first instar, only one feeding per day is sufficient but in the advanced stages, at least 2-3 feedings are required for proper growth of the larvae. Daily cleaning of the rearing bed and maintenance of hygienic conditions in the rearing room are desirable.

As the larvae are very sensitive to external stimuli, care should be taken to avoid any excessive handling or disturbance. The larva starts feeding from the margin of the leaf blade and proceeds towards the mid rib. Generally, it does not take up mid rib and petiole. While feeding, the larva holds the leaf blade or mid rib or leaf petiole with the help of claspers and abdominal legs. The average feeding duration in a stretch is about 30 minutes, but it does not extend beyond one hour.

The leaf requirement of young and mature larva differs markedly. On average, a first instar larva consumes only 0.21 g leaf in comparison to 60.33 g in final stage with total requirement of 72.83 g per larva.

Oak tasar larva thrives well between a temperature of 20 -25°C and can survive up to 30°C above which heavy larval mortality occurs. The early stages require higher humidity (80 -90%) than the advanced stages (70-80 %).

Young larvae relish green juicy leaves but tender leaves of pale green colour are injurious to health of larva. Early stages should be fed with tender leaves and advanced stages with medium mature leaves. The over matured or dry foliage should be avoided. Chawki rearing period is 12-15 days.

For late age rearing, proper space has to be given with proper ventilation. The late age requires 70-80 % humidity and proper cleaning of bed. The larvae should be fed four times in a day as 80 % of the food is consumed by the insect in late age rearing. Except, moulting stage, daily bed cleaning results in minimizing the loss of worms. Mature larva stops feeding and crawl for safe site for cocooning. These larvae are to be picked up and should be mounted in the bunch of hung twigs of food plants. The late stage larval period is 30 -35 days.

For indoor rearing, rearing house is the first requisite to protect the silkworm from direct sun light. The house should be dry, well ventilated and insect free. Before use, it should be properly disinfected with 5 % formalin. All equipments used for rearing should be disinfected.

PROCESSING OF TASAR COCOONS

Tasar Cocoon Sorting

Salient features of technology

- User friendly, easy to operate and occupies less space.
- Traditionally, the cocoons are separated manually without the use of any equipment.
- This invention shall help to separate the cocoons in a systematic manner, into A, B and C grade cocoons, with capacity to separate 40000 to 50000 numbers of Tasar cocoons in one hour time with 99 % accuracy.

CSTRI Automatic Tassar Cocoon Storing Machine

- The design and development of this machine is aimed at to overcome the drawbacks faced in the manual separation of the cocoons being practiced in the field.

Tasar Cocoon Cooking

Wet Reeling

Tasar cocoons cannot be cooked with plain water like Mulberry cocoons, the process of cooking / softening of tasar cocoons is entirely different which allows the cocoons to boil with chemicals and to facilitate for easy unwinding of filaments.

Open pan cooking

Cooking of tasar cocoons by using Hydrogen peroxide and soda ash method has been popularized by CSTRI specially for wet reeling. Central Silk Technological Research Institute has standardized this method of cooking, which gives better tenacity as well as elongation compared to traditional cooking. This cooking is particularly suitable for wet reeling.

Open pan cooking

Cooking process developed by CSTRI has two stages as explained below.

First stage: The tasar cocoons are taken in a metal cage, immersed in plain water and boiled for about 40 minutes. Care must be taken to ensure that the cocoons are not floating in the bath during entire process of cooking. Cocoons in the bath are allowed to cool for 30 minutes.

Second Stage: Cocoons are then transferred to the next vessel containing the chemicals dissolved in warm water (45°C – 60°C) for 20 minutes, depending upon shell weight with the following recipe:

Soda 8 gm/litre

Hydrogen peroxide 10 ml/ltr

Sodium Silicate 8 gm/litre

Cocoons are then deflossed and taken for wet reeling.

CSTRI Two in one Reeling cum Twisting Machine Silks

Salient features of technology:

CSTRI Two in one reeling cum twisting machine

- HDPE Reels have been provided to reel the Tasar and Muga cocoons without imparting twist to the reeled yarn. Reeling is done on front side of the machine.
- On the rear end of machine, twisting of reeled silk can be done, which can also be operated simultaneously.
- The production of the reeling machine is about 350 gms/8 hours.
- TPI of 7 to 12 turns per inch can be imparted on the machine as compared to the 5 to 9 in the earlier version.
- Suitable for both Dry & Wet Reeling to produce very fine denier yarn
- Machine designed ergonomically
- Reduces fatigue
- All machine controls easily reachable from the sitting position
- Energy optimization
- Suitable to twist single & 2ply yarn
- Suitable to produce both Warp & Weft Yarn
- Optimized to retain better Tensile and Cohesion properties

CSTRI motorized reeling cum twisting machine

Salient features of technology:

- Most popular and versatile reeling machine widely used in muga and tasar silk reeling.

- Reeling and twisting of reeled silk done simultaneously.
- The production of the reeling machine is about 250 gms/8 hours.
- TPI of 5 to 9 turns per inch can be imparted on the machine.
- Suitable for both Dry & Wet Reeling of Tasar yarn
- Suitable to twist 2ply yarn
- Suitable to produce Warp Yarn
- Ensures better earning potential to the Reeler in comparison with the traditional reeling methodologies.

Motorized Reeling cum Twisting Mahine

CSTRI solar operated spinning machine

Salient features of technology:

It is more environmentally friendly, economically beneficial and easily accessible. For easy spinning the flyer is mounted on ball bearing. The said machine is suitable for aged and handicapped persons and is also a low-cost spinning machine. The bobbin provided in the machine is easily removable and the hand spun yarn produced can be wound on re-reeling machine for making standard skein. The machine is provided with variable speed flyer and it is easy to operate.

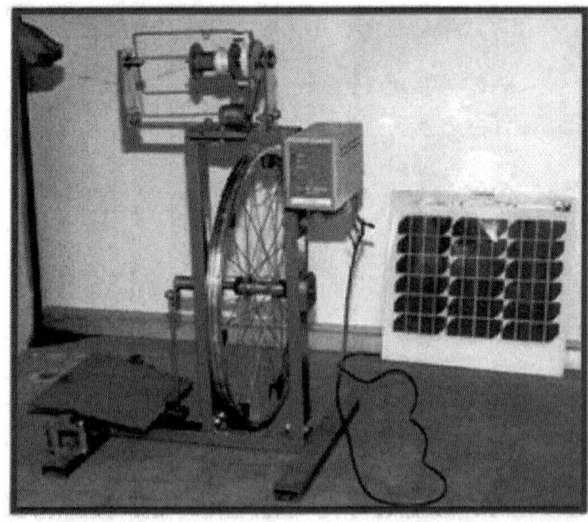

Solar energy operated spinning machine

DISEASES OF TASAR SILKWORM AND THEIR MANAGEMENT

Diseases of tropical tasar silkworm

Due to outdoor rearing practice, a considerable tasar crop is lost every year due to various diseases. The more common disease in tasar silkworm are **microsporidiosis, Virosis, Bacteriosis And Mycosis.**

Microsporidiosis

It is commonly known as pebrine disease. Pebrine in "Midi"language means " Pepper disease". The black spots resembling pepper sprinkled over the silkworm body.

Causal organism: ***Nosema mylittensis***

Symptoms

Until the disease is advanced, worms do not show any symptoms visible to the naked cye.

1. **Unequal size** of worms.
2. The worms become sluggish
3. Compared to healthy worms, affected worms are irregular in **moult** and are paler or more translucent
4. From III instar onwards, in severe case, pepper spots appear over the whole body of silkworm.

5. Worms lose their appetite.
6. Pupae infected with pebrine are light in weight and majority of them die.
7. Spins flimsy cocoons
8. Infected moths are generally deformed and have **crumpled wings**.
9. Infected adults lay unfertilized eggs.

Mode of transmission

Transmission of pathogen takes place through the following modes.

1. Pe roral
2. Transovum
3. Transovarial

1. Per oral

Infection takes place mainly **through food.** The larvae thus infected per – orally during **1st and 2nd instar show symptoms in late 4th and 5th instar and die before spinning.** If it is infected in **3rd instar the larvae show symptoms in late 5th instar** and die after cocooning.

Transovum: This type of infection occurs due to the contamination of egg chorion, a portion of which is eaten up by the larvae at the time of hatching.

Transovarial:

- The pathogen is transmitted along with the **egg cytoplasm**.
- As the yolk cell migrates to form the ovum in the ovary of the infected mother moth, the intermediary stages of the pathogen migrate along with yolk cells and get engulfed in whole egg.
- **Early laid eggs of infected moth may escape infection**, if the disease has not advanced, but those laid later show high rate of infection as most of yolk cells are affected towards the end of oviposition and as such mother moths harboring the disease transmit to the offspring's.

Detection

- A portion of eggs, egg shells and newly hatched larvae are grounded with **2% KOH** and entire suspension is centrifuged at **1000 rpm for 2min** allowed to settle and the sediment is examined under microscope.
- If pebrine is detected, the lot is **destroyed by burning.**
- During rearing, late moulter silkworm with **retarded growth, dead larvae and the faecal matters can be tested to detect pebrine.**

- In **pupal stage, detection of pebrine is not easy due to slow multiplication** of pathogen and hence examination of gut content and gonads are required to see the infection.

Control measure

1. Thermic control: Thermic control at **40^0C for 10 hrs**. to the infected eggs collected after 24hrs of oviposition is considerably effective in combating the disease.
2. Chemo control: Feeding of **Benomyl** a fungicide to the infected worms at **100-150 ppm** dose.
3. Administration of **0.1% Bengard and 0.005% carbestine/ carbendizim to the infected worms from II stage onwards is highly effective.**
4. **HCl treatment (sp.gr.of 1.05) to 72 hrs old egg at 40^0C for 10min.**

However, the best method is by disease free layings and elimination of infected ones.

Precautions

1. Maintaining hygienic condition
2. A thin layer of bleaching powder should be sprinkled under the bush and regular spraying **of 2% formalin.**
3. After the harvest, sample of seed zone should be microscopically examined and only infection free lots are retained in the grainage and the rest should be sent for reeling.
4. During egg laying, a thin layer of bleaching powder should be sprayed on the floor and then earthen cups / sweet boxes should be kept over it for egg laying.
5. The eggs should be first washed with **0.5% NaoH** and soap solution to remove **meconium** and then disinfected in HCl at 40^0C for 10min and washed with water.
6. During grainage/ incubation period, hygienic condition is maintained

Virosis

The virosis **or polyhedrosis** is virus disease of tropical tasar silkworm which is characterized by presence of large number of many-sided crystal polyhedral in tissues which contain virus particles within them.

A. mylitta is affected by cytoplasmic polyhedral bodies, multiplication of which occurs in the cytoplasm of the cells. The crop loss estimated is 20%.

Symptoms

1. Infected larvae grows slowly and greatly lag behind development
2. Disproportionately large **head or long** bristles
3. Later the polyhedra developing in the gut get frequently **regurgitated or voided** in large quantities with faeces.
4. In **6hrs** of infection, the feeding ceases and the larvae becomes immobile
5. In **12hrs**, body loses its natural shape, distends lengthwise and **turns brownish.** The larvae dies in 24hrs and on death, it **hangs** its head downwards remaining attached to the host twig with its claspers and **dark brown fluid oozes** out as drops from the mouth. The dead larvae produce **obnoxious odour** at this stage. In another 24hrs it becomes a mass **of melted** tissues.

Causative agent

It is caused by cytoplasmic polyhedral body which infects the cytoplasm of the cells in contrast to nuclear polyhedral virus. The size of polyhedra are usually **0.5 to 1.5m.** It looks **hexagonal** in shape, viruses are released when polyhedra is dissolved and the liberated virions are **spherical with 65nm** in diameter. The **double shadowing technique** has shown that CPV of *A. mylitta* is icosahedral.

Mode of transmission

It is transmitted **per orally and transovum**, high humidity with high temperature **predispose the disease.** The natural **epizootics** of disease is common in overcrowded population. The feeding of **juicy /tender** leaves to mature larvae induces polyhedrosis

Bacteriosis

The diseases of tasar silkworm caused by **pathogenic bacteria** are commonly known as bacteriosis. In tropical tasar, the crop loss due to bacteriosis is to the tune of **10-15%**. Tasar silkworm shows the **three** types of symptoms due to bacteriosis. They are

1. Sealing of anal lips
2. Chain type excreta
3. Rectal protrusion

Sealing of anal lips

- With the start of infection, the worm becomes **restless and irritable**.
- After 12hrs, larva becomes sluggish with **reduced feeding.**
- **In another** 12 hrs it excretes **soil coloured sticky semi fluid that seals the anal lips.**
- At this stage, larva stops feeding, becomes immobile and shrinks lengthwise.
- Ultimately larva dies and body colour becomes dark.
- Infected worms shows the presence of **Gram -ve - *Microccus* and Gram + ve *Bacillus*** in the form of rods in chains.

Chain type excreta

- **Sluggishness of the worm and slow feeding** are the primary symptoms of this bacterial disease.
- The worm becomes thin, long and soft.
- After 24 hrs it excretes out faecal beads embedded in **jelly like** substance in the form of chain that hangs down from the anal opening.
- The claspers lose their hold, body colour becomes slightly dark and larva stops feeding. At this stage larva does not respond to external stimuli and dies in 12hrs.
- Diseased worm shows species **of genus *Bacterium.***

Rectal Protrusion

- The infected larva becomes restless and reduces feeding.
- After 18 to 20 hrs. the rectum protrudes out from the anal opening in the form of **transparent bag filled with green haemolymph.**
- The anal lips dialate, feeding stops and body contracts lengthwise.
- In this condition, larvae remains attached to twig for 6-8hrs. and then falls down.
- The death occurs after 24hrs.
- Infected worms shows Gram-Ve *Micrococcus* and Gram +ve thin rods belonging to the genus *Bacillus*.

Control measures for virosis and bacteriosis

The normal route for both diseases in tasar silkworm is per oral and agro-climatic condition (high humidity and moderate to high temperature).

1. The ground area in and around should be kept neat.
2. Disinfection with 2% formalin or bleaching powder/lime in the field is necessary every 4 to 5 days.
3. Over matured, too tender and diseased leaves should be plucked off from the plantation.
4. Water logging in the field should be avoided which increases humidity.
5. Fresh solution of **0.01% sodium hypochlorite (2.5ml/lt)** should be sprayed on the worms when they are on the bushes once each in II, III and IV instar (24hrs after moult) and twice in V instar.
6. As an alternative TKO **(tasar keet oushadh) @ 1.5kg / 100 DFLs** can be dusted on the body of silkworm once in II, III and IV instar and twice in V instar.

Muscardine/ Mycosis

Tasar silkworm is found to suffer from stiffening of the body followed by fungus growth and subsequent mummification. The incidence is noticed in the mouth **of August to September and the crop loss is about 2-5%.**

Symptoms

- The infected larva becomes inactive and loses its apetite, the colour turns pale and body gets **hardened.** In about 12-14 hours, the larva hangs with its anterior or posterior half obliquely downward giving a characteristic dorsal bending.
- The worm at this stage looks very hard and pale and dies in 6-8hrs. Eight hours after death, the worm becomes **spongy and fragile**.
- In 16-18hrs, a white encrustation appears around each segmental ring and the larva gets more compressed laterally.
- After 24hrs, the **encrustations** covers the whole body. The dead worm becomes completely compressed laterally.
- **The white turns slightly green indicating formation of conidiospores.** The dead larvae becomes dry, brittle and mummified.
- Two types of fungi *Penicillium citrinum* and *Paecilomyces varioti* are the causative agents of mycosis.

Control measure

1. Spraying of **0.06% form**alin on silkworm, when there is an outbreak of disease.
2. Dusting TKO on silkworm before transfer.
3. Spraying **0.5% NaOH** on the worms I day after each moult.

MUGA SILKWORMS – BIODIVERSITY – DISTRIBUTION

In Eri and Muga silk production, India ranks first and second to China in Tasar silk. The popular items made from this silk are 'dhoti', 'chaddar', 'chapkan','pugree' and 'mekhala'. Commercial rearing is practiced in Sibsagar, Lakhimpur, Nowgong, Darrang and other districts. If the larvae are fed on mejankori leaves (*Litsaea citrata*)/ ***Antheraea assamensis*** **Helfer** produces mejankori silk.

This silk is very much admired for its durability, luster and creamy white shade. The muga reeling and weaving are done at Sualkuchi village. The natural golden color silk produced is known for its glossy fine texture and durability. The economic viability of muga industry needs no emphasis because of its cultural and traditional significance.

Besides this, with traditional skill coupled with the rich heritage of handloom and weaving of the region, the muga industry has offered a greater scope for generating employment for a sizable section of the rural masses in north- eastern India

Muga possesses unique characteristics such as unique method of cultivation and production, colour stability (everlasting), golden colour increases with each wash, tensile strength (4.53g/dn),strongest amongst all silks, UV absorption capacity (>80%),durability (over 50 years), acid resistant (resistant to concentrated Sulfuric acid) andthermal properties. Around 50,000 families directly engaged in muga farming.

The silkworm is multivoltine with two commercial crops during October-November and March-April. The commercial rearing operation is carried out in the outdoor conditions except cocoon formation and seed production activities.

The production process is highly influenced by climatic conditions and due to non-availability of hybrids, the production and productivity is static during last two decades.

Distribution

Assam accounts for more than 95% of the muga silk production. The culture is also spread in different districts neighbouring Assam in Meghalaya, Nagaland, Manipur, Mizoram, Arunachal Pradesh and West Bengal.

Muga silkworm, *A. assamensis*, occurs in the Brahmputra valley in Assam, East, West and South Garo hills of Meghalaya, Mokokchung, Tuensung, Kohima and Wokha districts of Nagaland, Lohit and Dibang valleys, Chanlang and Papumpare districts of Arunachal Pradesh, Tamenglong district of Manipur and Coochbehar district of West Bengal.

It also occurs in Northern Myanmar and the Kumaon and Kangra valleys in the western Himalayan hills, Sikkim, Himachal Pradesh, Uttar Pradesh, Gujarat, Pondicherry, Bangladesh, Indonesia and Sri Lanka Among the seven north- eastern states, muga production is confined, mainly to the state of Assam. Assam is the only state for production of reeling cocoons, whereas, other states have the privilege of producing major quantity of seed cocoons for commercial multiplication.

Assam produces 95 per cent of the total muga raw silk followed by Meghalaya. Contribution by other states is marginal. Muga rearing is considered profitable in upper assam districts, mainly in Lakhimpur, sibsagar, jorhat and dibrugarh.

These districts produce 90 % of total muga raw silk, however, the area under muga and other silkworm rearing activities being covered only to the extent of 10 per cent in these districts, there remains still a huge untapped potential. Thus, much more area in these districts can be brought under sericulture with an emphasis on muga silk.

Bio diversity of Muga

The muga silkworm is a single species with little genetic variation among populations, survives harsh climatic conditions and is subject to various diseases, pests and predators.

Host plants and biology

Bio types of muga

Although muga silkworm has not been successfully domesticated, attempts have recently been made to maintain it under semi-domesticated conditions and improve its economic traits . The main reason for this is the marked inbreeding and their hibernating over winter.

In particular different biotypes have been collected from various locations and maintained in culture.

Some muga biotypes, like Halflong green, yellow mutant, Kokrapohia green and wild hibernating type, were collected from different areas and kept in off-site conditions and hybridized in order to determine whether their offspring show hybrid vigour.

The biotypes were collected from Bhaktapara in lower Assam and Senchoa, Kamarbandha and Titabar in upper Assam, and wild muga silkworms, which diapause during winter, from a few areas in northeastern India.

Three biotypes of muga silkworm namely Sarubhagia, Barbhagia and Bor or Lebang are generally reported. Different colour forms were also collected from farmer's fields.

The muga silkworms collected during a survey conducted in the foothills of Arunachal Pradesh, Assam, Meghalaya and Nagaland were grouped into five biotypes:

- Type-W1: 24, Type-W2: 16, Type-W3: 12, Type-W4: 10 and Type-W5).

FOOD PLANTS OF MUGA SILKWORM AND MANAGEMENT

Som, ***Machilus bombycina***

Soalu, ***Litsaea polyantha***

1. Generally, Som is used for rearing of muga silkworm in upper Assam, while Soalu is used in lower Assam. Som tree is more prevalent in upper Assam and produces reeling cocoon whereas Soalu is more common in lower Assam and produces seed cocoon.
2. **Propagation of the host plants and their management:** The host plants of muga silkworms are available in nature. The plants are propagated in two ways (i) Sexual method and (ii) Asexual method. Som and Soalu plants are propagated through mainly seeds. Seeds are usually propagated by fallen excreta of birds with undigested seed scattered over a wide area. Seeds from selected plants ensure production of healthy seedlings.
3. **Propagation of Som:** Som can be propagated both sexual and asexual methods.

4. **Sexual Propagation:** Sexual propagation is through seedlings, particularly the seed propagation carries a varied population, this to utilize in selection and hybridization. For seed germination certain pre-requisites are needed to be fulfilled such as selection of quality seed, preparation of land, and the seed should be selected such that can definitely germinate. This is possible only when the seed is subjected to suitable environmental conditions, embryo of seed is alive, and healthy, in internal conditions of seed are favourable for germination.
5. The fresh seeds will have greater germination rate than the stored once. Seeds must be washed with fresh water until the flesh of fruit is withdrawn and dried well, however minimum moisture percentage should be maintained, i.e. at least 6 %. Sowing of seeds may be by way of broadcasting or sowing in lines.

Propagation through seeds: Propagation through seeds is one of the cheapest and easiest methods.

Merits: Som being a cross pollinated plant, sexual propagation introduces variability in the progeny and gives scope for selection of new varieties. It is suitable for large scale multiplication to build up stocks for preparation of grafts.

Demerits: Comparatively long gestation period to provide leaves for silkworm rearing. The desirable traits of improved cultivars cannot be perpetuated.

Source and time of collection of seeds: Seeds of Som become mature in April to June in Assam and other N.E. states. Collect mature seeds from the plants during April and May in lower Assam and May and June in upper Assam and other N.E. states.

Precautions

1. Collect seeds only from fully ripened fruits.
2. Do not damage seeds during extraction.

Viability of Som seed: There is no dormancy in Som seeds. Freshly harvested seeds have highest germinability. Seed is viable up to 10-20 days after harvesting of seed. The lose viability with the passage of time. The viability of Som seeds loses if preserved for a long period. Viability decreases after 20 days of storage till 45 days.

Selection of viable seeds: The seed quality can be selected by floatation test. Selection of healthy and viable seeds is done by floating the seeds in water. The healthy seeds sink and shriveled and unfertilized seeds float. The pulp of the seeds is washed off by kneading two or three times in running water and dried under shade for a few hours.

Storage of seed: The seeds may, however, be stored in moist seed beds for six to eight weeks under low temperature to prolong their viability. 10°C is the storage temperature of Som seed. It is desirable to predict the quality of the seeds in respect of germination capacity.

Raising of Som seedlings: Som seedlings can be raised by sowing seed in soil directly or in polytubes.

Pre-treatment of seeds and Seed dressing:

1. Prior to sowing, soak the seeds for 20-24 hours in water and treat the seeds with Bavistin or Carbendazim @ 2 g/kg seed.
2. For seed dressing Trichoderma pseudomonas @ 20 g/kg seed has been suggested in place of Bavistin.

Preparation of nursery Bed or Seed bed: Nursery can be prepared after monsoon from mature seed. Seedlings are available in plenty in a plantation before monsoon. Though seeds can be directly used for raising plantation, the present trend is it grow seedlings in the nursery and transplant them in the field to reduce the period of establishment and save wastage of precious seed material.

Layout of bed / Bed size

1. Select well drained high land in a shady place. The soil is dug or ploughed twice or thrice up to 30 cm depth, followed by leveling properly. Apply 6 cft FYM and mix thoroughly with the soil.
2. Seed beds of convenient size are prepared. Make 2 x 1 m beds and elevate the same up to 15 cm - 20 cm.

Seed sowing: Seeds are sown in lines in the prepared beds at a spacing of 15 cm in the row and 15 cm between rows at a depth of about 2-4 cm. Sow 2 kg seeds (approx.)/bed

Germination of Seed: Germination starts after 30-45 days of sowing. Germination of Som is 82% (average). Normally one seed gives rise to one seedling only but polyembryonate seedlings are also known to occur. In the polyembryonate seedlings, twins and triplets have been observed in Som which are rare occurrences. The frequencies of twins and triplets are 4.5% and 0.6% respectively. The twins and triplets after separation are capable of developing into independent plants.

Transplantation of seedlings to poly tubes:

1. The seedlings can also be raised in polythene tubes. Seedling of 6 months, 10-20 cm (3" to 4") height is collected for preparation of nursery. The seeds after collection are kept on moist bed under tree shade covered with moist gunny bags followed by sprinkling of water to maintain the moisture. The germination

starts after four weeks. The germinated seeds are sorted out every day and sown individually in polythene tubes of 9 x 6 inch size of 150-220 guaze, filled with rooting media and kept under tree shade.

2. Rooting media. A mixture of soil, sand and farm yard manure (FYM) in the ratio of 1:1:1 soil: sand: FYM is used as the rooting media for raising seedlings in the polythene tubes
3. Seedling survivability: 70-85(%)
4. **Gap filling:** Gap fills on the 10th day of sowing of seed on seed bed. Gap filling of seedling polytubes.

Short points- How you can grow muga food plantation

- Select a suitable site
- Prepare land
- Raise seedlings
- Plant seedlings
- Replenish nutrients
- Training and pruning
- Maintain the plantation
- Maintain chawki garden

Suitable host plants

- Muga silkworm is polyphagous in nature.
- Feeds on the foliage of a variety of food plants
- Primary food plants Som *(Persea bombycina)* and Soalu *(Litsaea polyantha,Juss)*

Secondary food plants are mentioned as below:

- Mezankari (*Litsaea citrate*)
- Kathalua (*Litsaea nitida*)
- Dighalati (*Litsaea salicifolia*)
- Patihonda (*Actinodaphne obovate*)
- Baghnala (*Actinodaphne anquistifolia*)
- Bogori (*Ziziphus jujube*)
- Bhimloti (*Celastrus monosperma*)

Select a suitable site

- Select high land for plantation which are well drained.

Prepare the land

- Prepare land by deep digging or ploughing up to depth of 30 cm depth.
- In the hilly tracts, terrace the land and raise bunds on all sides to prevent soil erosion.
- Mark out the site into convenient plots.
- Dig pits of one cubic foot size at a spacing of 3 m x 3 m.

Propagation through Seed

- Collect ripe and mature seeds during May-June.
- Select seeds weighing at least 0.3 g and measuring at least 0.7 mm in diameter.
- Remove the seed pulp by washing the seeds in running water.
- Test the seed viability by dipping seeds in water.
- Dry the seeds in shade and treat them with 0.2 % Ceresan.
- Mix the dugout soil of the pit thoroughly with 0.5 cft (5 kg) farm yard manure (FYM) and fill up the mixture in the pits.
- Sow the treated seeds after soaking in water for 24 hours.
- Prepare the nursery bed of 5x1.5 m size and 15-20 cm height.
- Apply 5 cft FYM and broadcast about 2 kg seeds in each bed and mulch with thatch.
- Seeds germinate in 45-60 days.
- Alternatively, sow the seed directly in polythene tube filled with soil, sand and FYM mixture (1:1:1).
- Transfer the 2-3 months old seedlings to 15 cm x 20 cm size polytubes
- When the seedlings attain a height of 30 cm, nip off the apical tips with 3 to 4 leaves by hand.
- Spray 0.03% Dimecron solution on the seedlings/
- Use seedlings for planting when they are 9 - 12 months old.

Propagation through single leaf and bud cutting

- Cut 4 cm long tender branches with single leaf and single bud of selected variety of som plants.
- Insert the cutting in a poly bag filled with coarse sand.
- The tube should be kept in shady place covering with polythene.
- Spray water to provide 90% moisture inside the cover
- Rooting started after 45 days and saplings are transferred to nutritional media when about 60% rooting is obtained.
- By adopting this methodology the gestation period of seedling for can be reduced.
- Use seedlings for planting when they are 9 - 12 months old.

Transplantation

- The monsoon season (April - August.) when the rainfall is high, is ideal for planting of muga food plants
- Select about 450 vigorously growing seedlings/ saplings for plantation in one acre of land.
- Transplant 9-12 months old seedlings or 6 months old saplings / layers/ stem cuttings into pits
- Plant one seedling/ sapling in each pit.
- Before transplanting, slit open the poly bag with a blade and remove it gently to avoid disturbance to the root system.
- Erect fencing around plantation site to avoid damage to plant from the cattle before plantation

Replenish nutrients

- Apply Farm yard manure (FYM) at the rate of 0.5 cft (5 kg) /plant/year and inorganic fertilizers @ 40g, 60g & 15 g NPK/plant /year in tw equal split dose up to 4th year
- Double the dosage from 5th year onward.
- Dig a circular trench of 6" to 9" deep around 60 cm circumference at the tree base and refill the same with FYM and NPK.
- Apply the inputs during March- April and September- October

Training and pruning of trees

- For better management and supervision of rearing and to obtain quality foliage plants are maintained at proper height and shape.
- This can be achieved by following proper training and pruning of the plants
- Periodically nip off apical tips of the plants at an interval of six months to get umbrella shaped crown.
- Remove unwanted and unhealthy branches.
- Train the plants by cutting side branches after 3 years of plantation.
- Resort to light annual pruning after 5 years of plantation
- Heavy pruning subsequently at intervals of 2 years.

Care during and after pruning

- The pruning cut should be oblique and anti-direction of sun rise.
- The bark around the cut should be intact without any peels.
- Minimise the injuries because injured regions are susceptible to various type of 'infection'.
- Apply parts of cowdung, mud and 0.1 per cent dursbin dust on cut ends.
- Apply FYM and fertilizer (NPK) doses immediate after pruning.

Pruning schedule

Crop	Pruning season	
	Early age worm	Late age worm
Aherua (Jun-Jul)	1st week of March	1st week of Feb.
Bhodia (Aug-Sept)	1st week of April	1st week of March
Katia (Oct-Nov)	1st & 2nd	week of July
	1st & 2nd	week of June
Jarua (Dec-Feb)	1st week of Sept	1st week of Aug
Chotua (Feb-Mar)	1st and 2nd	week of Oct.
	1st & 2nd	week of Sept.
Jethua (Apr-May)	1st week of Dec.	1st week of Nov

Maintain the plantation

- Weeds are a common feature in any cultivated land.
- Carryout regular weeding to avoid competition for nutrients.
- Remove weeds manually before they flower.
- Loose the soil to provide good aeration to roots and percolation of water.
- Control weeds like thatch grass with spray of glycel.
- Practice intercrops during gestation period in the vacant space of the plantation site.
- Intercropping control weeds and provides additional income on the other.

Maintain chawki garden

- About 20% plants should be maintained as 'chawki garden' by pruning at a crown height of 6 ft. above the ground.
- Prune/clip the branches of the chawki garden for chawki worms as per pruning schedule.

- Apply additional doses of FYM at 1cft (10 kg) and 80g ,120g, 30 g (NPK)per plant /crop after sprouting of new leaves in addition to normal dose.
- Ensure watering of the 'chawki plants' coinciding application of fertilizers and about 15 days before the brushing of silkworms in Chotua seed crop.
- Carryout regular weeding to reduce competition for nutrients and aeration in the soil.

Inter crops with Som

To increase soil fertility and soil health – dhaincha, sun hemp.

DISEASES OF MUGA FOOD PLANT SOM AND THEIR MANAGEMENT

Leaf Spot (*Phyllosticta perseae*)

Symptoms

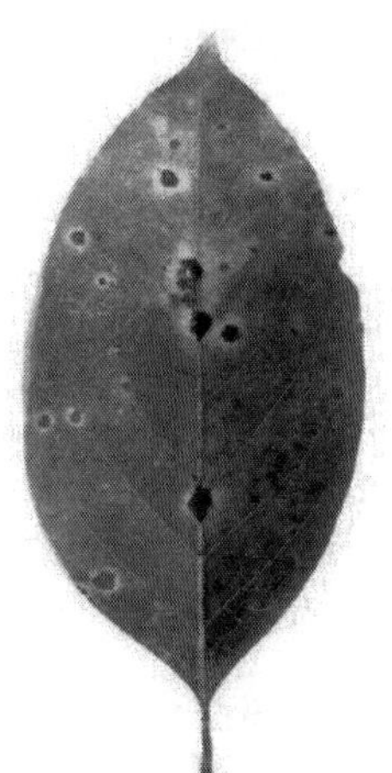

Leaf spot infected twig

- Appearance of circular or irregular brown spots surrounded by yellow margin on both surface of the young and matured leaves.
- More prominent on the upper surface.
- The first sign of the disease is appearance of slightly pale areas on upper surface of the leaves.
- The lower leaf surface shows collapse of epidermis, which loses contact with mesophyll tissues.
- As the disease progresses, the minute spots spread irregularly and become brown in colour, get collapsed and form larger patches causing drying up of entire lamina.
- Excessive spotting and destruction of green tissues of the leaf leads to heavy reduction in leaf yield.

Peak Season: July with 12-22 PDI.

- Foliar spray of 0.1% Dithane-M45 twice in 15 days interval controls the disease up to 85%

Leaf Blight or Anthracnose (Colletotrichum gloeosporioides Penz)

Symptoms

- Appearance of ash coloured, round to oval spots spread irregularly over the entire young as well as matured leaves.
- As the disease progresses, the spots get collapsed and malformed.
- Spots usually appear nearer to the leaf edges and infected area dries into brownish black colour.
- With the disease severety, brownish coloured lesions/ streaks (stromatic masses or sclerotia) appear on the twigs of the plant.
- The entire branch or top of the branch may get withered.
- Causes severe premature leaf fall.

Peak season:

- June- July with 48-60% plant infection and 14-22% leaf area destruction.

Management

- Pluck and burn the infected leaves.
- Since the pathogen is soil born and remain viable under the soil surface, practice deep hoeing to destroy the pathogen inocula.
- Cultural practices such as pruning and pollarding are effective in controlling the disease
- Spray 0.1% Indofil-M45 twice in 15 days interval.

Grey Blight (*Pestalotiopsis desiminata)*

Symptoms

- Appearance of small, oval brown grey lesions irregularly scattered on the leaves.
- Symptoms appear both in young and matured leaves.
- With the progress in disease infection, the spots get collapsed, malformed and ultimately the entire leaf dries up.

Peak season:

- May to August with 48-60% plant infection and 22-28% leaf area destruction.

Management:

- Spray 0.1% Bavistin on leaves twice in 15 days interval.

Red Rust (Cep*haleuros parasiticus* Karst)

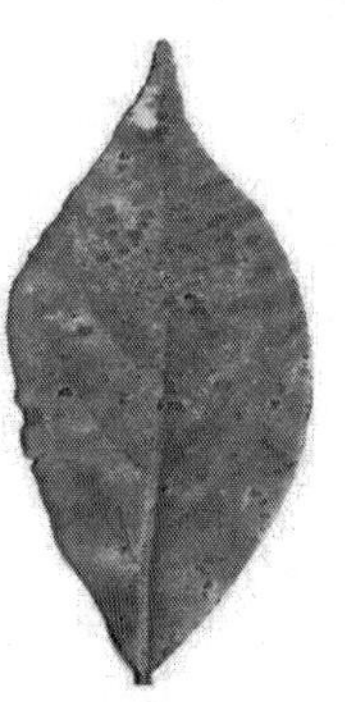

Symptoms

•Appearance of yellow green to orange or grey coloured hairy postules on the upper surface of the leaves.

- The postules tend to elongate into the stems and petioles.
- The affected branches become stunted and bear few leaves.

Peak season:

- May to August with 12-22% leaf damage.

Management

- Application of optimum dose of potasic fertilizers also helps in containing the disease.
- Pruning and plucking of infected plant parts and leaves help in reducing the disease spread.
- Spraying of 1 % Bordeoux mixture is general protective measure

DISEASES OF MUGA FOOD PLANT SOALU AND THEIR MANAGEMENT

Brown Blight (*Colletotrichum gloeosporioides* Penz.)

Symptoms: The symptoms appear roundish to irregular brown spots on young and mature leaves in the form of 'ash' colour, round to oval spots irregularly spread to the entire leaf.

- As the disease progresses, the spots, turning to grayish colour.
- The spots got collapsed and giving a blighted appearance.
- These spots usually appear nearer to the leaf edges.
- The infected areas dries up and become brown to black in colour.
- The top of the branch or the entire brunch may wither away

Peak season:

- Sept-Oct. is the peak season with 76 % plant infection and 42.47 % leaf area destruction.

Management:

- Plucking and burning of infected leaves at the initial stage can minimize the pathogen.

- Use of phyto-blighton (developed by CMERTI, Lahdoigarh) @ 50 ml/ liter of water during the month of June July to protects the leaves from the disease.
- Spraying of Copper oxiychloride (Fytolan or Blitox - 50) @ 3gm/liter of water can be used as a prophylactive measures to control the disease.

Red Rust (*Cephaleuros parasiticus* Karst)

Symptoms

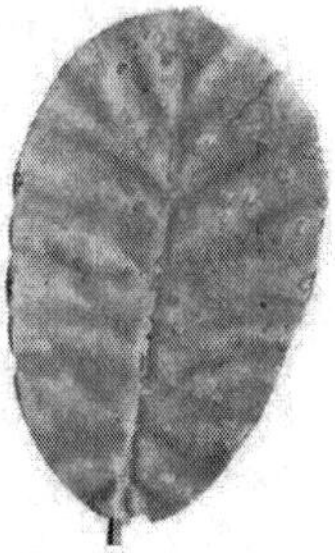

- Yellow green, orange or gray colour hairy spots appear mostly on the upper surface of the leaves
- The pustules are circular or irregular in shape and surrounded by chlorotic halos.
- The affected branches become stunted and bear fewer chlorotic leaves.

Peak season: May to August. Causes 39% leaf damage.

Management:

- Application of optimum dose of potasic fertilizers also helps in containing the disease.
- Pruning and plucking of infected plant parts and leaves help in reducing the disease spread.
- Spraying of 1 % Bordeoux mixture is general protective measure.

Stem Borer (*Zeuzera indica*)

Season: December- January

Nature of damage:

- The pest is found in both Som and Soalu plants and prevalent in entire north eastern region. The moths lay eggs on the bark of the tree and hatched larvae enter into the stem by making holes.
- The larvae feed on vascular tissues and pith. The life cycle takes nearly one year.

Extent of damage:

- Maximum infestation is 71% in Som and 73% in Soalu.
- The number of holes/ plants is 1 to 5.
- Size of the holes varies from 0.5 to 2.0 cm

Management

Biological method:

- Plugging of borer holes with cotton swab soaked in 5-15% plant extract of Pochotia/ Neem/Titabahak can be controlled up to level of 80% infestation.

Chemical method:

- Plug the borer holes with cotton swab soaked in 1.5% Nuvan solution followed by mud plastering. It controls up to 90% of infestation.

Leaf gall *(Pauropsylla beesoni* in Soalu and *Aspondylia sp* in Som plants)

Nature of damage: Gall is malignant tumour like growth on the leaves induced by the toxin like saliva secreted by gall insects inside the leaf epidermis.

Extent of damage:

- Maximum infestation is 30% in Som and 49% in Soalu.
- Size of the galls varies between 0.5 to 1.2 cm in diameter.
- Number of galls/ leaf ranges between 20 to 90.

Leaf gall of som **Leaf gall of soalu**

Extent of damage:

- September to November with78-80% infestation.

Management

- Pluck and burn the infested leaves.
- Cultural practices like pruning, pollarding, weeding and deep hoeing are also effective in controlling the pests.

Shoot Borer

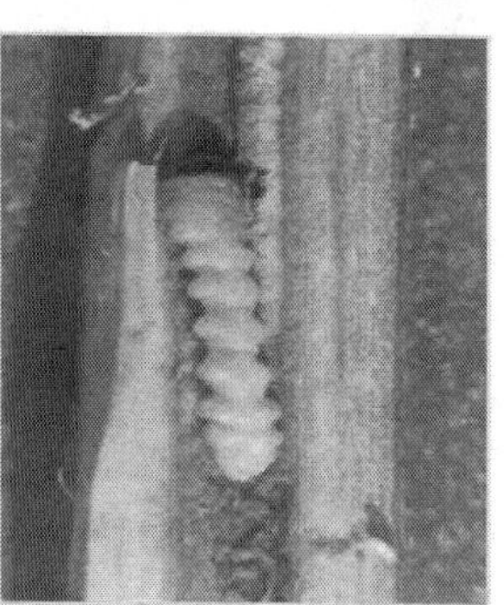

Nature of damage

- It is common in Som and found in apical twigs of the plants.
- Side shoots are bored from the nodal portion through the axils and the main veins of the leaves
- A tunnel is formed inside the main stem.
- Young plants die in severe cases of infestation.

Extent of damage: September to November with 70-80% occurs during

Management

- Mechanical control is the only way to reduce the infestation by cutting/clipping and burning the infested twigs.

Amphutukoni (*Cricula trifenestrata* Helf)

Nature of damage

- Feed on som leaves cause serious damage.

Extent of damage:

- June to September with 80-90% leaf loss.

Management

- Collect the cutterpiller in any stage of development and destroy.

MUGA SILKWORM REARING TECHNOLOGY

Muga crop cycle

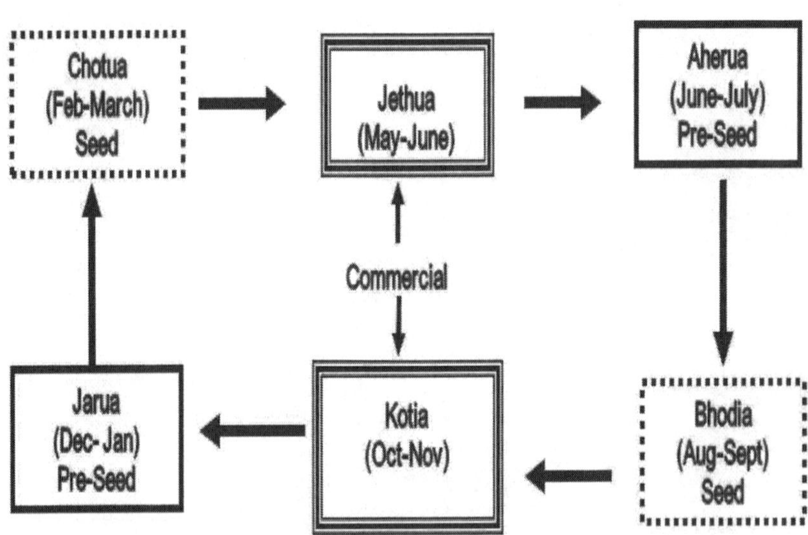

Climatic conditions during Rearing of Muga Silkworm

- During commercial crops the climatic condition mostly remain congenial with suitable foliage for Muga Silkworm rearing.
- During pre-seed and seed crops the climatic condition mostly remain unsuitable with high incidence of diseases and heavy infestation of pests and predators.
- Occurrence of heavy winds and hailstorms.

Importance of seed crop

- Successful rearing and harvesting of quality cocoons during commercial crops.
- Production of disease free and elite seed during commercial seasons.
- Thus, to utilise the full bearing capacity of food plants.
- To check the outbreak of disease which generally occurs due to rearing of inferior quality of seed.

Constraints of pre-seed and seed crop rearing

A. Jarua crop (Dec-Jan) & Chotua crop (Feb-Mar):

- Low temperature (15-25°C), low humidity (45-80%), short photoperiodism (6-8 hrs).
- Prolong larval period (40-45 days)
- High rate of uzi infestation (40-80%)
- High incidence of diseases *viz.* flacherie, grasserie and muscardine.
- Foliage less succulent, over matured and less nutritive.
- Poor cocoon quality.

B. Aherua crop (Jun-Jul) & Bhodia crop (Aug-Sept):

- High temperature (34-36°C), high humidity (81-91%), and their fluctuations.
- Water stagnation in rearing field leading to high humidity.
- Wastage of early stage worms due to heavy rain and hailstorms.
- High incidence of Pests and predators like ants, spiders, bugs, wasps, birds etc.
- High incidence of bacterial and viral diseases.

Imroved muga rearing package

A. pre-brushing care of food plants:

- Select systematic plantation in elevated plot for rearing.
- Maintain 20% plants as chawki plot by pruning at 6 ft height as per pruning schedule.
- Apply 1 cft FYM / plant and 40:60:15 gram NPK/plant in 2 split doses 1 week before pruning.
- Clear the undergrowths in the field.
- Remove dry, yellow, over mature, very tender leaves, dry twigs, ant and wasp nests from the plants 10-15 days before rearing.

- Resort to regular watering of plants 15-20 days before rearing to improve moisture content of foliage.
- Dust bleaching powder & lime in the rearing plot 1 week before rearing.

Pre-brushing care of eggs

- Sterilize the egg surface in 2% formalin followed by through washing in running water & dry in shade.
- Incubate eggs at 26 0 C & 85% relative humidity.
- Do not expose the eggs to heat, bright light or chemicals.
- Adopt prescribed methods for transportation of eggs.

Care during brushing & chowki rearing

- Cover the rearing plot with nylon net to protect silkworm.
- Brush the newly hatched eggs between 5-7 am on tender leaves.
- Consider the worms hatched on the 1st-3rd days only for rearing.
- Brush only 2-3 dfls per plant so that the worms can feed till the 3rd instar.
- Keep regular vigil on the movement of pests & predators at the rearing site during brushing

- Brush worms indoor under inclement weather conditions.
- During rearing apply grease on the trunk & cover with polythene sheet to prevent the worms crawling down and crawling up of ants on the tree.

Care during late stage rearing:

- Dust bleaching powder & lime in the rearing plot 1 week before rearing for transfer of worms.
- Tie polythene barrier around the trunk.
- Use nylon nets to protect the worms from pests & predators especially during Jarua and Chotua crops.
- Clip tender leaves along with worms when there is still 20% foliage on plants to avoid starvation.
- Avoid frequent handling of worms.
- Use only disinfected chaloni.

Collection of ripe worms, cocoon harvesting

- Collect ripe worms in bamboo baskets (khasa)
- Allow the worms to spin cocoons on mountage (jali/bamboo box mountage)

- Hang jali in semi dark, well aerated and rat proof room for better cocooning.
- Harvest cocoons only after pupation (8-10 days)
- Sort out good, flimsy, uzi infested cocoons after harvesting.
- Cocoons weighing 5-6 gram with 0.4-0.5 gram shell weight are ideal.

Prophylactic measures:

- Collect dead/diseased/irregular/weak worms & litters and examine under microscope.
- In case of pebrine out break, destroy the worms and defoliate the plants followed spaying 1% formalin solution in the rearing plots and avoid rearing in the plot for 6 months.
- Collect irregular, weak, diseased and dead worms in basin with 2% formalin solution and burn/burry in pit away from rearing.
- After completion of rearing, resort to light clipping of the plants & apply FYM/NPK.
- Clean the rearing plot and dispose the unwanted remains in the compost pit.

DO'S AND DON'TS FOR PRE-SEED & SEED CROPS

DO'S	DON'TS
JARUA & CHOTUA CROP ➢ Maintain 20% plants as chawki plot & prun the plants at 6 ft height 4 months before rearing season. ➢ For late age rearing, prune the plants 5 months prior to rearing. ➢ Apply FYM & NPK one week after pruning. ➢ Resort to regular watering of plants to improve moisture content of foliage. ➢ Remove dry, yellow, over mature, very tender leaves, dry twigs, ant & wasp nets and spider webs from plants. ➢ Dust bleaching powder and lime mixture (1:9) in the rearing plot 1 week before rearing. ➢ Cover the rearing site with nylon net. ➢ Brush the worms indoor under inclement weather conditions i.e. hailstorms, heavy rain, whirl winds etc. and delay brushing by 2-5 days. ➢ Allow the worms to spin cocoons on mountage in semi-dark, well aerated and rat-proof room for better cocooning. ➢ Harvest cocoons only after 9-10 days of spinning. ➢ Sort out good, flimsy, uzi infested cocoons after harvesting.	➢ Do not brush excess worms / tree. ➢ Do not transfer the worms during moult. ➢ Do not allow dead / diseased/ weak worms to remain on food plants to prevent spreading of disease. ➢ Do not prepare Jali with wet leaves as it may attract fungus. ➢ Do not remove the under growth of rearing site completely to conserve moisture among the rearing plants. ➢ Do not mount healthy, ripe worms along with uzi-infested worms. ➢ Do not harvest cocoons before complete pupation. ➢ Do not disturb the worms while spinning.
AHERUA & BHODIA CROP ➢ Select systematic plantation in elevated plot for rearing. ➢ Maintain 20% plants as chawki plot by pruning at 6 ft height 3 months before rearing season. ➢ For late age rearing, prune the plants 4 months prior to rearing. ➢ Apply FYM & NPK to the plant one week after pruning. ➢ Remove dry, yellow, over mature, very tender leaves, dry twigs, ant & wasp nets and spider webs from plants 10-15 days before rearing.	➢ Do not select plantation of low lying areas as rearing site. ➢ Do not brush worms on trees having over matured foliage. ➢ Do not brush worms towards direct sunlight. ➢ Do not conduct rearing in un-maintained plots. ➢ Do not brush excess worms / tree.

[Table Contd.

DO'S	DON'TS
➢ Apply commercial grease on tree trunk to prevent entry of ants from soil to plants.	➢ Do not transfer the worms during moult.
➢ Dust bleaching powder and lime mixture (1:9) in the plot 1 week before rearing.	➢ Do not allow dead / diseased/ weak worms to remain on food plants to prevent spreading of disease.
➢ Brush the newly hatched worms between 5- 7 AM on tender leaves opposite to sunlight.	➢ Do not prepare Jali with wet leaves as it may attract fungus.
➢ Transfer the worms when there is still 20% foliage on plants to avoid starvation.	➢ Do not disturb the worms while spinning.
➢ Use only disinfected Chaloni.	➢ Do not harvest cocoons before complete pupation.
➢ Cocoonage hall should semi-dark, well aerated and rat-proof.	
➢ Harvest cocoons only after 6-7 days of spinning.	

36

GRAINAGE TECHNIQUES IN MUGA EGG PRODUCTION

Seed production is the most important aspect of Muga silk Industry and the success of rearing mainly depends on it. Seed production may be divided into four divisions, viz.

(A) Selection of seed cocoons,

(B) Preparation of seeds,

(C) Seed examination and

(D) Seed supply.

The seed preparation houses should be spacious. They are to be provided with a microscope, heater, lights, air cooler, humidifier, incubator, refrigerator, sprayer including formalin, etc. They should be free from rats, lizards, spiders, ants, etc.

Important characters for selection of Muga are viability, fecundity, larval period, larval weight, mortality, disease resistance, cocoon weight, shell weight, filament length, denier, etc. Selection may be done considering all required characters of the Muga silkworm (chromosome number n=15) throughout its life cycle.

SELECTION OF SEED COCOONS

Better seed cocoons are always required to produce better silkworm seeds. If possible, it is better to select them during the larval stage. There are hibernating and non-hibernating Muga strains. We should try to increase the production of hibernating Muga broods.

Experienced Muga rearers visit the Muga rearing in the last stage before maturity. They observe the worms, the mode of their eating leaves on the tree, visual examination of some of the worms, percentage of mortality, etc.

Healthy worms usually eat the leaves from the top of a tree and even the twigs. They are bright green in colour. 2-3 litters are present at the rectal portion of a healthy worm. They show their sensitiveness without vomiting upto 4-5 times, while touched.

Seed cocoons are usually selected and graded on visual examination through long practical experience, after the harvest.

Seed cocoons should be selected from and around "Bhorpok"(mid-maturity) stage to obtain approximately equal number of male and female moths for seed preparation. Healthy, well built, robust cocoons with live pupae, fine denier silk, more percentage of silk content, good reelability, longer filament length, compact in texture, should be selected. It is better to carry seed cocoons along with the 'Jali' to avoid damage during transportation.

It is advisable to collect healthy disease-free seed cocoons from a distant place so that change of food, climate, etc, occurs and better cocoon crop may be obtained.

Selected seed cocoons are to be kept in rat-proof well ventilated rooms in wire-netted and bamboo cages (chakaripera) in one layer.

More space is required for preservation of Muga seed cocoons. Seed cocoons may be kept in garland when a large number of Commercial seeds are to prepared. Each garland should have 50-150 cocoons.

Seed cocoons received from different areas are to be kept separately Seed cocoon should not be stored in damp, poorly ventilated and in dark places. Wire-netted cages are to be used to maintain the purity of the strains of Muga, especially where parent stocks are maintained. The cages may be arranged in tiers for economy of space and protected from ants.

Preservation of spring and autumn breed's seed cocoons should be in cold storage for delay and regular emergence of moths. Muga undergoes diapause in pupal stage. This will also help in controlling pebrine disease. Such preservation of seed cocoons will help in avoiding summer and winter crops. This will ensure the seed supply in autumn and spring rearing. It is better if the seed cocoons may be preserved under regulated temperature and humidity followed by proper incubation for uniform emergence.

Seed Cocoons may be preserved in cold storage at 25°C first and then reduce the temperature to 5°C after every 12 hours up to 5°C. Seed Cocoons may be preserved upto 4 months. Similarly, while releasing these preserved seed cocoons, the temperature should gradually be increased by 5°C after every 12 hours and then to kept at the room temperature. Seed cocoons also may be preserved in high altitude during summer with proper care in transportation.

Necessary steps are to be taken to control the life-cycle of Muga silk-worm by artificial means subjecting the pupae to hibernate for avoiding the summer and the winter rearing. It may be induced by controlling the photoperiod at the larval stage.

The photoperiod effect combined with temperature and humidity renders some Muga-worms to go for diapause and others to continue with the voltinism.

The price of seed cocoons is too high but the rearers purchasing capacity is limited. So, marketing of seed cocoons should be regulated.

Rearers co-operatives may also be formed to avoid exploitation by middlemen. Introduction of a crash programme for production of more muga seed cocoons would also ease the position.

PREPARATION of SEEDS

Muga seed preparation is time consuming and hazardous task. Usually four cocoons are required for preparation of a single laying.

Moths emerge from the seed cocoons after 2-4 weeks from the date of cocooning according to season. In normal condition, the emergence of moth takes place from evening till midnight.

Male moths emerge earlier than the female moths. They can easily be identified. The males have broader antenna and narrow and small abdomen. The females have smaller antenna and large abdomen.

Seeds cocoons are to be refrigerated at 5-10°C in early pupal stage and incubated at 34°C to synchronize the time of emergence of moths of different strains of Muga for hybridization. Moths may be refrigerated for 3-4 days without adverse effect.

Male moths may also be refrigerated for about a week. They may be utilized for a second time pairing, if and when necessary.

Refrigeration for a longer duration is not advisable. Moths are to be selected after emergence considering their health; natural brown colour with thick wings, without any deformity, and urine like that of pure milk, etc.

The male and female moths are allowed to pair in the cage naturally. Some of them may not pair for which mechanical means, such as , moving them by hand and blowing air by mouth for early pairing are to be adopted by the breeder.

Maximum care must be taken to avoid damage to the female moths. Unpaired female moths may also be kept outside in a safe place tying them on "Kharika" for pairing with wild males after darkness in the evening and should be collected in the early morning of the following day.

Moths prefer darkness for pairing with comparatively lower temperature and high humidity. Male moths are more active and smart fliers. Female moths are passive and generally do not fly.

The female moths are tied by one of the hind wings and fastened to a "Kharika" first and then allowed them to pair. Paired moths are taken out from the cage carefully and the female ones are tied, unpaired and kept for egg laying by hanging the "kharika" and strings/wire arranged already for the purpose.

Moths are usually allowed to pair for about 12 hours due to practical reason though 4-5 hours are sufficient for fertilization. They start pairing in the evening and are unpaired in the following morning by hand. One must take care at the time of unpairing so that female moths will not be injured.

Application of light or heat will help in separating couples by themselves. A single male moth can be used for a second coupling without affecting fertility and egg laying when male moths are less. After unpairing, the female moths are kept for egg laying on "Kharikas". They prefer darkness for egg laying like that of other satuniids.

They start laying eggs within 12 hours. They are allowed to lay eggs for 3 days, which are considered for rearing purpose. Eggs laid latter are less viable and the worms become weak, and development is poor.

A female moth usually lays 150-300 eggs. Muga moths lay maximum eggs in spring and autumn, and minimum during summer. Eggs adhere to "Kharika" or in a small bamboo/paper basket with cover for basic seed preparation to help in proper examination.

More moths may be allowed to lay egg together in longer bamboo Kharikas for large scale seed production. The moths with Kharika should not be exposed to sunlight and rain. Likewise, more unpaired female moths may be kept in paper/ bamboo basket for preparation of large scale loose eggs but it is not advisable as it would be difficult to examine the mother moths. Sometimes smoke or light is to be introduced in the room as shock treatment to the female moths.

It gives better effect in eggs laying. Deposition and preservation of eggs should be conducted in a cool room with sufficient humidity.

Incubation may be practiced for regular and uniform hatching at 40°C. It will also help in controlling pebrine disease. The moth is a non-feeding stage and dies within 7-15 days after emergence.

Continuous inbreeding will lead to degeneration of the merits of the race and also reduces vigour. Hybridization (out-breeding) in between two different group of Muga is essential for commercial rearing. Domesticated Muga may be hybridized with the hibernating wild Muga from time to time to give them hybrid vigour and to avoid the troublesome summer and winter broods.

Muga silkworm is a primitive insect and has no means of large scale hybridization in absence of related strains. Government farms should maintain parent stocks of different strains collecting from various places to study their behaviours and inter or intra-specific hybridization for keeping hybrid vigour. It would help to prepare hybrid seeds in subsequent generations.

Survey of potential seed producing areas and collection of several strains from different area are necessary. These may be reared in small batches in a farm and all the cocoons harvested from such rearing may be mixed allowing them to have random mating which will give hybrid vigour. An alternative to this, these strains may be reared separately in different farms and only male cocoons are to be exchanged to prepare hybrid seeds among them.

Seed examination

Visual examination play an important role in seed examination one may select better layings through his long practical experience. Eggs with glue bigger in size, laid in clusters with fibrous body dust and brown in colour are said to be healthy.

Weak and dead moths irregular layings, less number of eggs in a laying, under developed and unfertilized eggs, etc, are to be rejected before seed examination. This is to be followed especially for parent stock maintenance. It will also help in quick microscopic examination. It is advisable to conduct examination in all the stages of the life-cycle of the Muga silkworm.

Microscope, glass slides with cover, mortar and pestle, scissors, caustic potash solution, etc., are essential for seed examination.

After egg laying for three nights, the abdominal portion of mother moths are cut by scissor and crushed with the help of mortar and pestles, sprinkling a few drops of distilled water or ½ drop of 2 percent caustic potash solution. Then a drop of the suspension is placed on glass slide by stick/rod covered with glass cover and examined under microscope. The eye-piece, objective, etc. of the microscope should properly be cleaned and accurate focusing must be there for correct microscopic examination.

It is better to examine a mother moth at a time and in case of basic seeds it is a must. In case of large scale industrial seed production, examination of ten moths taking at random from a particular lot is to be examined at a time.

The organisms that cause diseases of Muga silkworms are virus, bacteria, fungus and sporozoa. Generally, the eggs laid by moths having mild viral, bacteria or fungal infections may be use for rearing as there is difficulty to get sufficient seed cocoons. But we should not consider any when they are infested by sporozoa (pebrine), which is transmitted from mother to the off-springs, through eggs.

Such infected mother moths and their eggs must be destroyed without delay by burning and burying them at a distant place. Loose eggs are to be collected on the fourth day of oviposition after examination of mother moths and washed (disinfected) with 2 percent formalin.

Washing of eggs for about 5 minutes with 2 percent formaldehyde solution followed by washing in clear cold water and drying in shade is essential to avoid contamination of diseases before supplying to rearers.

Heat treatment for pebrinised pupae and eggs may be done like that of oak tasar after proper experimentation. Disinfection with Kharika is more advantageous. Disinfected eggs are to be kept without overcrowding in cool and moist place.

Hatching may be delayed for few days by refrigerating the eggs if and when necessary after 48 hours of oviposition. But is should not exceed more than a week for normal hatching.

Loose eggs after disinfection may be preserved in a box having an arrangement for light on the top and for air on the sides. Eggs in boxes may be incubated at a required temperature and humidity which will help in uniform hatching. The eggs inside the box should be kept in one layer and not in heaps.

Seed supply

The process of transportation of seeds has a direct effect in the cocoons crops. No abnormal heat and direct sun should be experienced by the eggs during transit. Otherwise, it may result the failure in rearing, especially as it causes mortality in the 5th instar before maturity.

One must realize that the embryo is being developed inside the egg like that of the development of the baby in the mother's womb. Seeds should reach the destination within the seventh day of oviposition.

A well-perforated wooden box with a thin cloth lining inside is to be used for sending seeds (loose-eggs) with proper operation. It is advisable to transport eggs on 4th/5th day of incubation.

Quantity of eggs is to be put inside the box according to its capacity and in no case it should be over loaded to avoid damage to developing embryos. Eggs may be supplied along with "Kharikas" for nearby places.

Eggs are to be transported preferably in cooler hours. Tender leaves may be put in the eggs carrying boxes, if hatched eggs are transported. It is better to supply seed cocoons with "Jali" instead of seeds wherever and whenever possible with proper care in transportation. Seed cocoons should be transported carefully after 6-9 days of spinning.

Government should take appropriate steps to supply muga seed cocoons and disease free seeds to the rearers at reasonable cost in required seasons. Rearers usually cannot continue the rearing for more than three generations of a particular lot of muga due to successive inbreeding. Field Staff should arrange for microscopic examination of mother moths in rearers' house to get disease free eggs.

Short points about Grainage of Muga silkworm

Muga Silkworm Grainage Technology

Introduction

- Quality seed production is one of the most important activities in Sericulture.
- Rearing of healthy seed is prime requisite for success of a seed crop.
- It is highly essential to adopt the prescribed procedure meticulously for production of silkworm seed cocoons and seed (eggs).

Grainage operation

- Grainage is the site of production of disease-free laying (Dfls).
- It is therefore imperative to adpot strict hygienic methods for production of disease-free laying.
- The grainage building should be well aerated with cross ventilation and high roof with false ceiling.
- A broad range of temperature between 22- 28 °C and 75-85 % relative humidity are congenial for grainage operation.

Method of disinfection

- Seal the grainage hall 5-7 days prior to consignment of seed cocoons.
- Drench the walls & other bigger appliances with 5% bleaching powder solution (add 50 gm bleaching powder per litre water) or 2% formalin solution (Formalin 1 part: 18 parts of water).
- Dip smaller appliances in 5% bleaching powder solution.
- Use high grade bleaching powder with 30% chlorine content.
- Prepare fresh disinfectants by mixing 5 gram slaked lime / liter of 2% formalin solution (prepare 2% formalin solution and add 5 gm slaked lime per litre).
- Carry out disinfections perfectly on sunny days.
- Spray 2% formalin and 0.5 % slaked lime mixture @ 1liter per 2.5 sq. mtr and leave the hall closed.
- During high humid condition fumigate the hall with 5% formaldehyde solution for 24 hrs at least 3-4 days prior to and immediately after grainage.
- Protective measures to be adopted like gas mask, hand gloves should be used during disinfections.

Seed cocoons collection

- Select only fully formed & compact cocoons from Bhorpak (mid period of ripen worms) or plus-minus 1 day.
- If needed transport the seed cocoons in single layer on trays or cushioning with straws to avoid jerking. Avoid exposure to direct sun light, rain etc.
- Consign the cocoon on trays / cages in single layer
- Store the seed cocoons in single layer.
- Maintain temp. & humidity at optimum level.

Norms for selection of seed cocoons

Crops	Larval wt. (gm)	Fecundity (no)	Wt. of eggs/dfl (gm)
Aherua (P2)	6.90<	145<	0.90<
Bhodia (P1)	7.23<	160<	1.00<
Aghenua(P2)	6.40<	130<	1.11<
Chatua (P1)	7.00<	150<	0.90<

Moth Emergence

- Place the emerged moth in mating cage at 2:1ratio (2 male and 1 female) and place the cage in dark and well aerated place.
- Keep excess male moth under Nylon net for future use in well aerated, cool and shady place.

Moth Coupling

- Complete the moth pairing during evening hours.
- Allow 5-6 hours of coupling.
- During shortage of male moth, allow he same male moth a rest for 4 hours after first pairing and then use it for 2nd pairing.
- Natural coupling is preferred.
- Attend decoupling during the night itself.

Oviposition

- Put the fertilized female moth in Oviposition bag, hung on aerated place.
- Allow 3 days for oviposition.

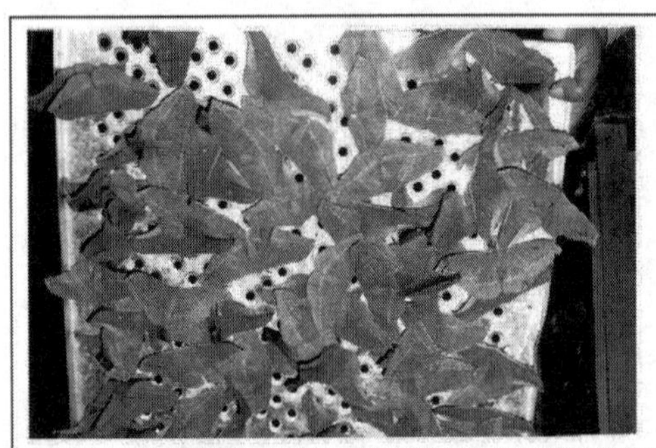

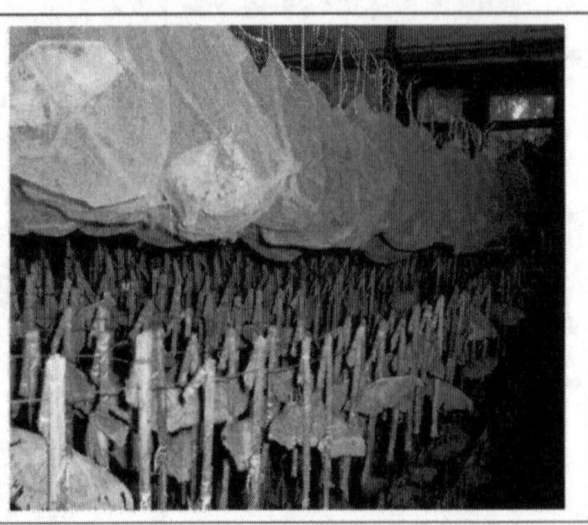

Moth Examination-(Fujiwara method):

- Mark the mother moth serially after 3 days oviposition for examination.
- Dissect the entire (other than head and wings) portion of abdomen and crush in moth crushing set with 0.08% potassium carbonate solution.
- Filter the solution with the help of thin layer of cotton spread over the funnel and poured the filtrate in to 15 ml tube.
- The tubes containing the filtrate are allowed to centrifuge for 4-5 minutes at 4000-5000 rpm for precipitation.
- Precipitation in the tube will be observed after removing the water from the tube.
- Put a small smear of the precipitation with the help of a small and thin stick on a slide and examine under microscope with 15 x 40 magnification.
- Sort out pebrinized laying, collect and burn.

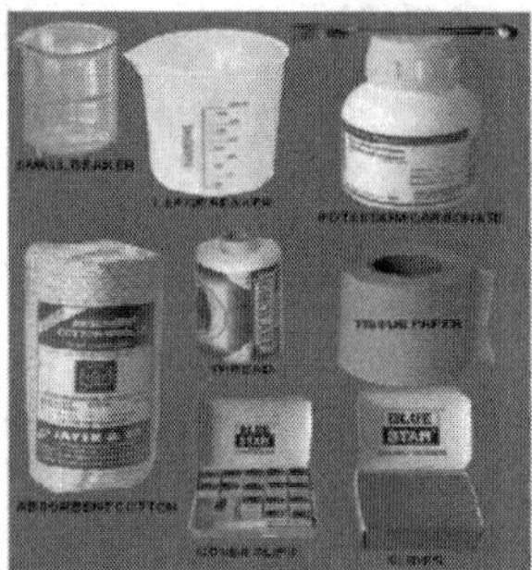

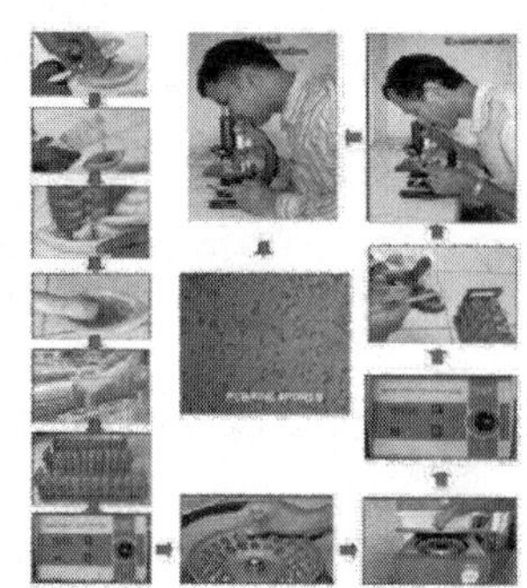

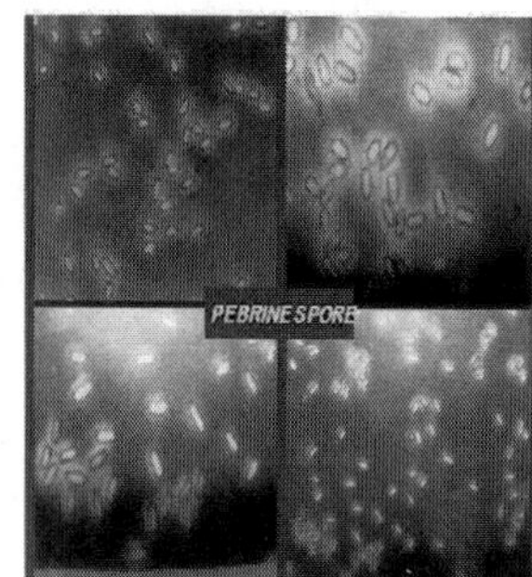

Egg Surface Sterilization, Drying and supply

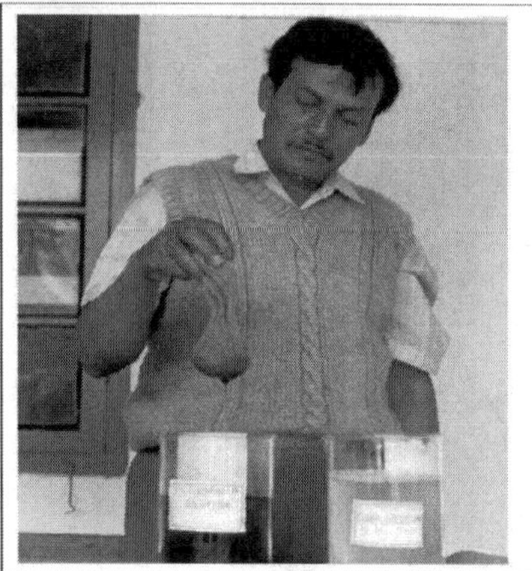

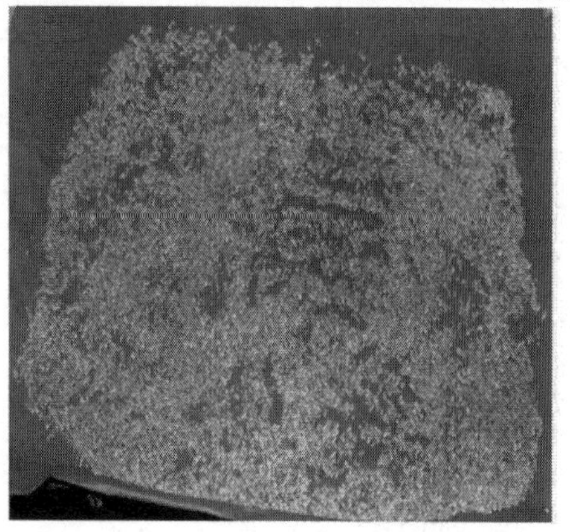

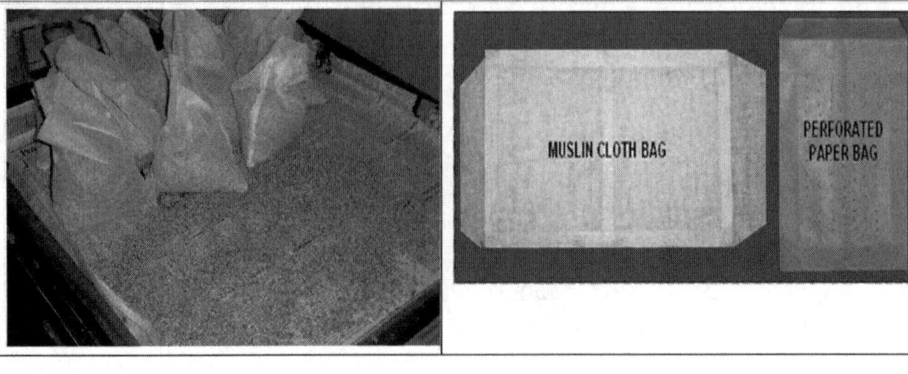

- Wash the dfls in 2% formalin solution (1 part formalin: 18 part water) for 5 minutes and then in running water till disappearance of formalin smell.
- At high temperature above 30°C, 2-3 minutes washing in 2% formalin is adequate.
- Place the egg in a single layer on blotting paper and dry in shade in case of loose eggs.

Egg Incubation

- Put the eggs in BOD incubator at room temperature.
- Adjust BOD temperature to 26° C +- 1 °C in ascending/descending order allowing the eggs to remain at least 12 hrs at each step to avoid temperature shock.
- Maintain BOD RH% at 80-90 % by placing saturated solution of Potassium Chloride in a wide petridish.
- Incubate till hatching.
- Restrict temperature fluctuation during incubation.
- In case of non-availability of incubator the incubation room should be maintained at 24-28° C and with relative humidity 75-90% through indigenous technology.

DO'S	DON'S
➢ Disinfect the grainage hall & equipments 3-4 days prior to grainage operation.	➢ Do not wash the grainage hall with plain water prior to spraying of formalin, bleaching powder solution.
➢ Prepare fresh disinfectant and use immediately for better result.	➢ Do not conduct grainage operation without prior disinfection.

[Table Contd.

Contd. Table]

DO'S	DON'S
➢ Spinkle Bleaching powder with Slaked lime mixture (100 gram beaching powder + 900 gram slake lime) at the entrance and around the grainage hall @ 200 gram per sq, mt in every alternate day. ➢ Use disinfectant musk and gloves during disinfections. ➢ Maintain proper aeration in the grainage hall. ➢ Maintain temperature 26-300C & 80-90 % RH during cocoon storage & grainage operation. ➢ Store seed cocoons in a single layer in moth cages to avoid disease contamination. ➢ Attend decoupling after 6-7 hrs of coupling. ➢ Upon detection of pebrine burn the infected moths along with the eggs. ➢ Use soak pit away from Grainage hall. ➢ Dispose off crushed moths etc with 2% formalin in soak pits away from grainage hall. ➢ Burn of dead / decaying / infected pupae / moths / eggs and melted cocoons.	➢ Do not store seed cocoons in a heap. ➢ Do not store cut/pierced cocoons, dead moth etc in the grainage hall. ➢ Avoid dampening of grainage hall. ➢ Do not enter the grainage hall without disinfecting hands/feet. ➢ Do not transport seed cocoons/ eggs during sunny hours of the day. ➢ Do not use poly bags for carrying eggs. ➢ Do not allow unauthorized person to frequent the grainage hall. ➢ Do not use poly bags for carrying eggs. ➢ Do not allow unauthorized person to frequent the grainage hall.

MORPHOLOGY OF MUGA SILKWORM

- Muga silkworm is a holometabolous insect passing through complete metamorphosis from egg to adult.
- The life cycle lasts for 50 days in summer and max. 150 days in winter.
- The silkworms are reared out door and as such are exposed to vagaries of nature.

Stages	Days required	
	Summer	Winter
Egg	7	15
Larval	24	52
Spinning stage	3	6
Pupal	14	44
Moth	2	3
Total :	50	120

Egg

The eggs are brownish, streakless and is 2.8 x 2.5 mm in size and 9 mg in weight. The follicular imprints consist of a single pattern with oval main cells.

Larva

- The newly hatched larva is characterized by prominent black inter segmental markings.
- The body colour is yellowish with brown head.

- It measures 7 x 1 mm and weights 7 mg.
- The tubercles are yellow and provided with setae.
- LP 3-4 days in summer, 6-8 days in winter.

II instar

- The body colour is green and the head is with brown colour.
- At maturity, the length is 9 x 2 mm weighing 15 g .
- Tubercles are blue coloured. LP 3-5 days in summer, 7-10 days in winter.

III instar

- Two prominent rectangular black marks are seen on prothoracic hood which is replaced by a pair of semilunar deep brown markings in the subsequent instars.
- LP 5-7 in summer, 10-13 days in Winter.
- Tubercles are violet in colour.
- The anal flap bears posterior a rectangular black mark.

IV instar

- U shaped markings are seen in IV instar with the two arms joining the lateral line.
- Prominent small black eye spots seen on lateral side of hood.
- Tubercles are red.
- Larval period: 7-10 days in summer, 12-15 days in winter.

V instar

- 4 cm to 5.5 cm and weighs 4.12 g - 5.21 g.
- Tubercles are brick red.
- U shaped marking are seen on the anal flap with a black inner and a deep brown outer border. Each segment bears a pair of dorsal tubercles, a pair of upper lateral, a pair of lower lateral tubercle.
- LP 10-12 days in summer, 16-19 in winter. Wt 10-15 g.
- The area above the plans has rectangular black marking early in the third instar, which subsequently assumes the shape of a thin black band.
- The lateral line is cream coloured in the first instar. Later in third instar turns yellow in colour. In the terminal portion, a faint chocolate colour line is present above the lateral line.

Pupa

On maturity, larva, spins cocoons after selecting a suitable site for pupation. The larva prefer space between two to three dry or semi dry leaves. The pupa is copper brown and measures 3.2 x 1.8 cm and weights 5.7 g.

Spinning period is 3 days in summer and 7 days in winter.

Pupal period: Summer 14 days, Winter 40 days

Sex differentiation

Pupa has a fine longitudinal line on the eighth abdominal segment in female and is absent in males.

Cocoon:

Single shelled, light brown, oblong, closed, reelable and slightly flossy with a weak peduncle.

- Size 5.2 x 2.4 cm,
- Weight 6.3 g,
- Shell weight 0.5 g,
- Shell ratio 9.5%
- The cocoon is golden brown or glossy white.

Moth

The fore and hind wings of moths are brown rarely with a pinkish tinge. The ante median line (AM) and oblique line (OC) have a white border on the inner surface.

	Male	**Female**
Body length	3.0 cm	3.5 cm
Wing expansion	13.0 cm	15 cm
Wing area		
Forewing	1662 mm^2	1857 mm^2
Hind wing	1181 mm^2	1351 mm^2.
Post Median line (PM)		
	Bordered by a single White lining on either side	White inner and Pink outer line
Ocellus		
	In forewing ocellus Measures 14 mm^2 and in hind Wing 25 mm^2	Forewing ocellus measures 23 mm^2 and in hind wing it is 35 mm^2

Table 1. Morphological characters of muga silkworm (*A. assamensis*).

Sl.no	Characters	Wild stocks	Cultivated (semi-domesticated)
1	Voltinism	Bivoltine / Multivoltine	Multivoltine
2.	Number of moults	4/3	4
3.	Food plant	Soalu / Dighloti	Som / Soalu
4.	Incubation period (days)	8–12	8–10
5.	Body colour	Black with yellow streak, Green	Black with yellow streak, Green
6.	Head colour	Light black	Brown
7.	Egg size (L × B) mm	2.5–3.0 × 2.0–2.5	2.1–2.8 × 2.0–2.4
8.	Egg weight (mg)	8. 36–9.33	5.83–9.83
9.	Larval duration (days)	22–50	22–45
10	Mature worm weight (g)	9.0–14.0	8.5–13.5
11	Cocoon colour	Light golden	Light golden
12	Cocoon shape	Elliptical	Elliptical
13	Cocoon weight (g)	4.5–8.55	2.90–7.70
14	Shell weight (g)	0.60–0.96	0.18–0.65
15	Filament length (m)	410–506	126–398
16	Filament fineness (denier)	4–6	4–6
17	Percentage of silk reelable	40–46	37–63
18	Silk recovery (%)	42–55	40–42
19	Fecundity (number of eggs)	235	173

REARING OF MUGA SILKWORM, *ANTHERAEA ASSAMENSIS*

Among the seven north- eastern states, muga production is confined, mainly to the state of Assam. Assam is the only state for production of reeling cocoons, whereas, other states have the privilege of producing major quantity of seed cocoons for commercial multiplication.

Assam produces 95 per cent of the total muga raw silk followed by Meghalaya. Contribution by other states is marginal. Muga rearing is considered profitable in upper assam districts, mainly in Lakhimpur, sibsagar, jorhat and dibrugarh.

These districts produce 90 % of total muga raw silk, however, the area under muga and other silkworm rearing activities being covered only to the extent of 10 per cent in these districts, there remains still a huge untapped potential. Thus, much more area in these districts can be brought under sericulture with an emphasis on muga silk.

ESSENTIAL ITEMS

The essentials items for Muga rearing are as follows:

- Bamboo baskets,
- Mounting brushes (Kharikas),
- Secateur,
- Butterfly catching nets,
- Bow with pallets,
- Bird scaring instruments,
- Nylon mosquito nets,

- Wire-netted cages or Bamboo cages (chakaripera),
- Triangular bamboo shieves (chalani)
- Alluminium pans
- Bamboo poles,
- Gamaxin,
- Tugon bait,
- Gum,
- Lali, molasses,
- Dry leaf Jalis, etc.

All rearing appliances must be disinfected with 4 percent formal dehyde solution in time.

Muga silkworm (*Antheraea assamensis*; family Saturniidae) is semi-domesticated and multivoltine. It is possible to rear 4-6 crops in a year. The rearing is out-door and exposes to environment conditions and attack of parasites, predators and diseases.

It results in heavy losses due to weather conditions, pests, diseases and natural calamities. These losses are due mainly to the faulty selection of rearing places (Muga chung) and food plants, improper handling of worms, poor supervision of rearing, intensity of natural enemies, etc.

Smoke from various engines and workshops, dust from gravel roads, some of the pesticides and herbicides used in agricultural fields and tea estates, D.D.T. spraying for eradication of Malaria are also some of the factors hampering the healthy growth of Muga-culture.

REARING PLACE

A suitable rearing place (Muga chung) is to be selected. It should not be low lying and shady. Medium size full-grown Som/Soulu trees with more branches and fresh leaves are to be selected for rearing.

Early stage worms should preferably be reared on dwarf plants for reducing loss of worm. Tall trees will reduce the production. It is also difficult to hang "Kharikas" and to transfer the worms to other trees after exhaustion of leaves on tall trees. Worms also find it difficult to crawl up the tall trees in search of suitable leaves. It increases the mortality of worms of course, flies wasps etc., may not be able to attack the worms easily in tall trees as they can fly to a particular level only.

The rearing place (Muga chung) should be cleaned including the base of all trees to avoid pests. The dry and diseased branches, twigs and leaves, nests of ants and wasps, etc, are also to be removed before starting the rearing.

Plants infected with ants, aphids, wasps, termites are to be avoided for rearing. A band of straw should be wrapped around the trunk of each tree at a height of about one meter. This will act as a barrier for the worms from coming down and prevents ants, etc., from going up. Banana or pine apple leaves may be put above this band so that worms cannot come down after exhaustion of leaves. These will facilitate to collect them easily.

Ashes, lime and sand mixed with kerosene may be put at the base of trees. Gamaxin may also be sprinkled at the base of trees to prevent ants but it must be done at least before a fortnight of mounting the worms. Fish refuse, mollases etc., may be kept at the base of trees to attract the ants.

Ants should be burnt regularly in the morning and evening. The nests of ants/wasps from nearby bushes/grasses are also to be cleared. It is better if the apical buds in the branches may be removed before two weeks of rearing as buds and every tender leaves are not suitable for rearing. Insecticides/fungicides should not be sprayed before rearing.

Hatching

- The eggs of muga silkworms are brownish in colour.
- They hatch out within 1 to 12 days of oviposition during summer and 16 to 25 days in winter. The worms start hatching from early in the morning till about 10:00 a.m. under normal condition. Hatching will be over within 24 hours.
- Newly hatched worms eat their egg shells.
- Worms hatched out during first two days are to be considered for farm rearing as they grow uniformly and are healthy.
- Larvae hatched up to 4 days are to be kept for commercial rearing.
- The percentage of hatching is 40% to 95 according to the season.
- Hatching will be delayed during summer by keeping the eggs at 6°C to 10°C and hastened during winter by exposing them to 28°C to 30°C as experimented at Muga Research Station, Dhakuakhana (Assam). Eggs kept in controlled condition and incubated eggs will hatch uniformly and worms would complete maturing with 3-4 days.

- Eggs should be kept in perforated boxes in an disinfected room and ants, lizard, etc. should not be there.
- Eggs with kharikas (bamboo orthatch) or the loose eggs in bamboo/paper baskets are to be tied with or hang on suitable leafy branches and on the trunk when the majority of the worms are hatched out (4 A.M. to 10 A.M.).

Mounting of hatched worms

The newly hatched tiny worms will crawl up by themselves to the leaves of Som/ Soulu by their natural instinct. They start eating from the margin of suitable leaves. They eat up the midrib and even the petiole of the leaf during the last two instars.

Another method of mounting the newly hatched Muga worms is that small twigs bearing soft and tender leaves are lightly put ever the newly hatched worms on the "Kharikas" or in the egg boxes/baskets. They will crawl up to the leaves.

The twigs with the worms are then tied with suitable branches at different places of the Som/Suala trees for uniform distribution. They are to be put either on the east or north side of the trees to get sufficient sunshine by the help of bamboo poles.

The suitable time for mounting the newly hatched worms is morning. It is not advisable to mount during heavy showers, heavy wind or storm. It is better if a limited number of worms can be mounted on a particular tree, where leaves will be sufficient upto maturing to avoid frequent transfer of worms. 5 layings may be a reared in a full grown Som/Soulu tree and about 200 cocoons may be harvested in one crop.

Overcrowding by mounting more worms on a tree by the rearers compel them to eat hard leaves unsuited to their age. It is the main cause of flacherie desease of Muga. It is desirable to mount the worms hatching in different days on separate plants.

Kharikas/bamboo baskets are to be collected, disinfected with 2% formaldehyde solution and preserved properly for future use. The younger worms prefer soft and tender leaves and matured leaves at the advance instars stages. It has been observed that the rate of feeding is higher at night than in the day time.

They take rest (about 5 minutes) after every stretch of feeding. Healthy worms take the whole leaf even soft twigs from the top of the tree. Larvae move much during the first instar-1 stage in search of the suitable leaves than during the other instars. If and when change of food plant is necessary, worms should be mounted first on Sualu and then transferred to Som. Worms find difficult to eat Sualu leaves due to presence of gall wasps and of flowers. Likewise, fly pests are more in Digloti rearing and the fruits of Digloti also hamper the worms to a great extent.

Moulting

Muga worms are green in colour with thorny warts. Worms move more before moulting to find out proper place for shelter.

Moulting period varies from 24 to 36 hours depending upon this season of rearing. The larvae eat their cast-off skin. No disturbance and transfer of worms should be there during moulting.

The transfer of worms after exhaustion of leaves should be done after 3-4 hours of moulting (ecdvsis). Day and night constant and close watching is indispensable to avoid pests and to scare birds, bats, etc. This will also help to assess the condition of the worms.

Dead worms should be collected twice daily in the morning and evening. They are to be burnt and buried away from the rearing place. Microscopic examination of dead and diseased worms should be done to detect diseases. It is advisable to conduct rearing of unhealthy and weak worms separately on fresh plants.

Muga worm start descending of their food plants when leaves are exhausted. They are picked up, sorted out according to their stages and hung on few leafy branches of other suitable tress with the help of triangular bamboo shieves (chalani) and bamboo poles.

Another method for transferring the worms is keeping leafy twigs along with the worms may be put on other suitable trees with the help of bamboo poles. One should handle them carefully to avoid injury to worms by forcible detachment. Worm falling from the trees should be picked up by some suitable leaves (branches).

Worms defecate orange shaped litters with six longitudinal furrows. Worms usually hide on the under surface of the leaves and have a strong gripping power. They are very sensitive and disturb easily.

They stop eating leaf and shrink their body (expressing their anger) when disturbed by movement, sound or touch etc. If removed by force, they vomit a liquid. Rough surface (bark) of the trunk of the tree should be made smooth. It will help in picking up worms easily without damage during transfer. Smaller worms would be picked up first and the bigger worms later. The blue colour larvae are bigger and heavier than the normal green colour larvae.

Larval period

The larval period varies from 22 to 50 days according to the season of rearing. At the end of fifth instar, the worm attains maturity, stops feeding and empties its alimentary canal by passing out the last excreta (green semi-solid) mass followed by coloured slimy substance).

Worms consume about 20 per cent leaves during first instars and 80 percent during the last instar alone. Matured Muga worms become slightly smaller, flabby and translucent and make a hallow sound on touch.

They come down from the tree in the evening due to effect of negative geotropism. It makes the collection of worms easier for the rearers. They are collected by the rearers in bamboo baskets quickly and may be put on 'Jalis' directly. The effective rate of rearing varies from 20-60 Cocoons per laying, depending on the season rearing.

Feeding

Muga worms feed on Mejankari and champaka leaves produce creamy white silk. It fetches a higher price. Of course, the growth of Mugaworm on Mejankari is said to be slow and unhealthy in comparison to that of som Plants. However, rearing on Majankar may be conducted where it is growing abundantly for producing creamy-white 'Mejankari' silk.

Maturing

Generally, a crop completes maturing within 3-4 days. Of course, some crops take upto two weeks to complete maturing. The worms maturing earlier produce male moths, those that mature later produce more female moths and worms maturing in between (Bhor-pok days) produce an approximately equal number of male and female moths.

The matured worms are then taken to the seed preparation house or any other suitable place and put them into the Jali prepared earlier from suitable dry leafy twigs and kept hanging vertically in dry place for spinning cocoons after being counted. The number of worms in a 'Jali' are to be adjusted according to the space available in it. 'Jalis' may be arranged on bamboo mats where matured. The mature worms are then taken to the seed preparation house or any other suitable place and put them into the Jali prepared earlier from suitable dry leafy twigs and kept hanging variety in dry place for spinning cocoons after being counted.

The number of worms in a 'Jali' are to be adjusted according to the space available in it. 'Jalis' may be arranged on bamboo mats where matured worms are kept and they will crawl up to the 'Jalis' easily. Someone should observe the worms atleast for an hour after hanging the 'Jalis' as some worms way fall down from the 'Jalis' while searching for suitable place for cocooning.

Those worms are to be kept on 'Jalis' again. Sex in larval stage can easily be distinguished by the external genital markings. Four dots in case of female and a V-shaped marking the in male worms are seen in between the 8th and 9th segments on central side. Sex markings in pupa are more prominent than in worms.

Separate cocoonage should be used for males and females in case of production. 'Jalis' are to be prepared from the leave of suitable trees. The quality, size, colour, etc of the cocoons differ according to the food plants on which they feed. It is said that the age of the tree has a great effect on the colour of the silk produced. The cocoon spun by Muga worms feeding in Diglati leaves is smaller to that of other Muga food Plants.

Cocooning

- Cocooning is completed within 3-4 days in summer and 7 days in winter.
- Pupation is completed after 15-30 days of cocooning.
- No disturbance should be there to the Jalis up to a week during cocoons formation. The female worms spin bigger cocoons than the males. Usually cocoon formation starts during the day time. Cocoons are harvested normally after a week of spinning during summer. Of course, it is delayed up to 1½ weeks during winter.
- The normal and the inferior cocoons should be harvested separately. Cocoons are cleaned of the adhering leaves at the time of harvesting.
- Good and healthy cocoons are kept for seed production, and the others for reeling and spinning purposes. The weight of cocoons, silk shell, pupa and the length of filament varies according to season.

Indoor rearing

The indoor rearing for the whole larval period may be practiced for producing basic seed cocoons and for maintaining the parent stocks. It may be done by bringing leafy branches or Som/Soulu trees and putting them in some container (noncorrosive) having cold water.

These branches are to be changed from time to time whenever necessary to keep the freshness of the leaves. It should be done after proper experimentation for better harvesting. It may result about 95 per cent success of the cocoon crop. It does not require regular watching like that of outdoor rearing, reduces expenditure, loss of worms and mortality.

It also protects the Muga worms from their enemies like ants, wasps, birds, bats, etc. Continuous indoor rearing for many generations throughout the larval period may weaken the strain. So, indoor rearing up to second moult or third stage and outdoor rearing for the remaining instars preferably under net covering with proper care will help to maintain the health and vigour of the race intact and in harvesting more cocoons.

Muga worms like to take water drops accumulated on the leaf for which sprinkling of water once a day on the twigs is advisable in case of indoor rearing. Windows should be kept open from time to time as required. Disinfection is to be done to the rearing houses used for indoor rearing of Muga worms and also the rearing appliances like that of Eri and Mulberry silkworm rearing houses. Chawki rearing especially during summer may be adopted for better rearing results.

It is not possible to conduct large scale rearing of Muga worms indoor. The outdoor rearing for the whole larval period also increases loses to the commercial rearers. So, after rearing up to third instar indoor, worms should be transferred to dwarf leafy trees for outdoor rearing, for the remaining larval period. Dwarf plants covered with fine nylon mosquito nets or Wire-netted cages for outdoor rearing of young worms at least up to the second moult will also save the worms to a great extent from ants, wasps, flies, etc.

Worms hatched out from the eggs of different mother moths are to be reared on separate trees without mixing them Cocoons harvested from such rearing should be kept separately. Male and female moths emerging from these cocoons are to be allowed to pair disallowing brother sister mating (inbreeding). It facilitates in keeping the hybrid vigour in subsequent generations. It will help in the maintenance of the purity of the Strain without deterioration. Further research is necessary in evolving and developing some suitable interspecific hybrid for the greater interest of the Muga silk industry.

Yellow body mutant of Maga should also be tried for outbreeding which would help in maintaining vigour and health of the worms. The body weight of Muga worms is abouts 0.008 gms to 5.5 gms from 1st instar to 5th instar. The weight of cocoon varies from 4.5-6.0. gms and the shall weight from 0.30 gms to 0.60 gms.

Season of rearing

The rearing of two major commercial crops. (in four batches) a years may suitably be practiced during spring (early and late) for producing reeling cocoons. They may be adjusted according to the local conditions. It is better to conduct rearing of summer crop in cold climate (at high altitude) as mortality and flies, wasps, etc, are more during summer in hot places. It will help in supplying the seed cocoons for the autumn crop. It will also provide hill amelioration to the worms.

Maximum possible layings should be reared during springs (first crop and autumn third crop) rearing as these are the two most suitable seasons for rearing of Muga. It will increase the quantity and the quality of cocoons. Necessary steps must be taken to control different diseases and to get rid of various ants, wasps, flies, birds, bats, etc., for harvesting a successful cocoon crop. Tugon bait, gum, lali, molasses, etc. may be used in alluminium pans for controlling and killing ants, wasps, flies, etc., Birds scared away by bows and pallets and other bird scaring instruments.

Most of the wild Muga worms (yellow, blue and orange in colour) are bivoltine and do not emerge during winter. The collection of wild Muga seed cocoons from different places of hills and plains of Meghalaya and Assam and also from certain places will help the rearers in improving the production of quality cocoons by hybridizing them with semi-domesticated strains. It is essential to multiply the hibernating strains of Muga worms as such breed will eliminate the hazardous winter rearing.

They are vigorous and stronger and will solve the problem of seed cocoons to a great extent. Rearing of selected strains may be conducted in suitable green/ glass houses (specially early spring crop when hail storm, pests are prevailing), under controlled temperature (24°C-30°C) humidity (75%-85%) and light. Suitable selected clonal plants on tubs may be used for the purpose for making nutritious leaves available. This may also help in reducing degeneration of Muga worms. Rearers usually do not select the trees in advance before taking up rearing for which leaf quality varies and affects the growth of Muga worms.

PROCESSING OF MUGA COCOONS

Processing of Cocoons

- Muga cocoon are formed with continuous filament of 350-550m. For production of 1kg of muga raw silk, 4500-5500 cocoons are required.
- The existing traditional muga reeling device is "BHIR", where two persons are required to produce 60-80g of raw silk per day.
- Hence, CSTRI muga reeling machine for production of warp yarn and "BANI" reeling for welt yarn are recommended for muga cocoon reeling.

Selection of reeling cocoons

- Select compact and well formed cocoons.
- Deformed, double and flimsy cocoons are not suitable for reeling.
- Cocoons of commercial crops (Kotia and Jethua) are suitable for reeling.

Stifling, drying and storage of cocoons

- Stifle the cocoons immediately after harvest.
- Sun drying and smoking are two traditional cocoon stifling methods, where drying of cocoons is uneven leading to deterioration of cocoons within short time.
- Stifling in hot air oven or Ushnakuti is the recommended method as it takes care of both stifling and drying.

- Put the cocoons in hot air drier/ Ushnakuti at 95 oC and lower the temperature of the drier gradually to 55 OC for a period of 7-8 hour for the whole process of stifling and drying.
- Keep the stifling and well drier cocoons in well aerated cage made of iron wire mesh to avoid damage by insect pests and predators.
- Cocoon storing room should be dry and well ventilated.
- Dust slaked lime and bleaching powder to keep the room damp proof.

Cooking

- Cook the cocoons in boiling alkaline solution in open pan.
- Take 3 litres of water in an aluminum vessel and boil.
- Add 9-12g g soda and stir well. When foams are raised, put 200 cocoons in the boiling solution and boil for 5-10 minutes depending on quality of the cocoons.

Deflossing and reeling

- Remove the uppermost-entangled flossy layer of individual cocoon shell for taking out continuous filament end.
- During deflossing care should be taken to take out the mass first from lateral side and not from anterior or posterior side of the cocoon.
- Maintain basin temperature at 40-50 0C for wet reeling.
- Since single cocoon filament denier is 4.5-5.5, feeding of cocoons should be according to the requirement of the warp and weft yarn denier (35-40 / 50-55).
- After reeling of first batch, change the basin water and go for the second batch.

Re-reeling

- For hank making, improved re-reeling device with fixed circumference with diamond crossing mechanism may be used.
- The machine should be run at a uniform speed for unwinding of filament at uniform tension to suitably stretch out the filament laid on cocoons shell in a shape like “8”.

Weaving

- Fly shuttle loom is generally used for muga weaving. For designs, jacquard/ dobby mechanism is incorporated.
- The preliminary operations include sizing of warp yarn and warping.
- Some weavers dye the muga yarn with direct dyes to make the colour uniform.
- The warping is done in horizontal drum or in hand reel.
- A fly shuttle loom can consume 2-3 kg warp yarn and 4-5 kg welt yarn for one setting. Production in fly shuttle loom is 2-5 meters/day depending on type of fabric to be woven.

PESTS AND DISEASES OF MUGA SILKWORM AND THEIR MANAGEMENT

Pests

Uzi fly (*Blepharipa zebina* and *Exorista sorbillans*)

It is the major pest of muga silkworm. It is prevalent throughout the year attaining the peak during December to March. The fly lays eggs on the integument of the worms in the dorsal and dorso-lateral side.

After hatching from the eggs, the maggots of the fly penetrate into the larval body and feed on the tissue of the worm. The mature maggots come out of the larvae/pupae and undergo pupation in the rearing field or grainage hall. The uzi infested muga silkworm dies during larval or pupal stage.

Control

- Rear the silkworm under nylon mosquito net during peak infestation period (December to March), which ensures 80-90% control.

- During transfer of late stage worms, remove the fly eggs from the integument of the silkworm larvae with the help of forceps.
- Keep the rearing field clean and dust with bleaching powder during rearing.
- Mount uzi infested worms in separate 'Jali'.
- Harvest and stifle the uzi infested cocoons on 4th or 5th day of spinning.
- Collect and destroy the maggots/pupae of the fly.
- Burn the heavily infested worms.

Apanteles:

- *Apanteles glomeratus* (Larval parasitoid) usually infects the early stage silkworms. It is prevalent during summer and winter months of the year.
- *Apanteles* lays eggs inside larval body of the silkworm by inserting the ovipositor through tubercles. The maggots of the fly feed on the tissue of the silkworm and come out through the tubercles after maturation. The mature maggots form pupae in aggregation outside the body of the silkworm larvae.

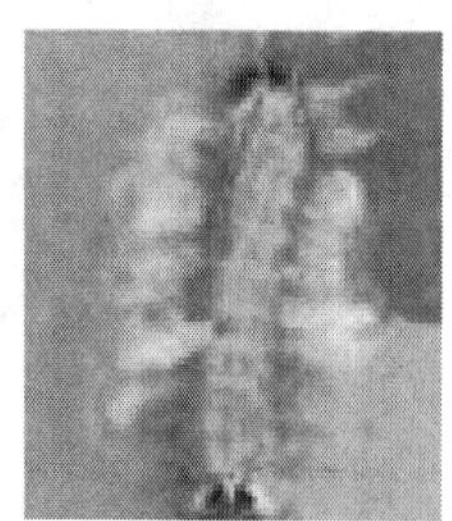

Control :

- Rearing of silkworm under nylon mosquito net prevents *Apanteles* infestation.
- Keep the rearing field clean and dust with bleaching powder during rearing.
- Collect and destroy the maggots/pupae of the fly along with the silkworm larvae.
- Ants, Wasps, Hornets, Cantheconid bugs, Preying mantids, Birds, Owls, Bats, Monkeys, etc., are the major predators of muga silkworm.
- Rearing under nylon mosquito net and mechanical control are the only methods to prevent the attack of predators.

Wasp (*Vespa orientalis*)

- It occurs during June-July to August-September months.
- It attacks early stage worms by lacerating and picking young age worms.
- It can be controlled by covering silkworm rearing by nylon nets and destroying hives.

Red ants

- The red ants are also serious pest in many muga growing areas. It attacks 1st stage worms. The lost due to red ants are reported to be 5-10%.
- They can be controlled with the spray of 2% Rogor (insecticide) before 15 days of rearing or burning down their nest well before the rearing.

Grass hopper:

- They attack the 2nd to 3rd stage worms.
- Lost due to grass hoppers are minimal

Diseases of muga silkworm

Pebrine: Pebrine is the most serious disease of muga silkworm caused by a protozoan of *Nosema* sp. It is unique in being transmitted to offspring by transovarial/transovum means from mother moth.

Occurrence: The disease may occur in all seasons of the year.

Symptoms:

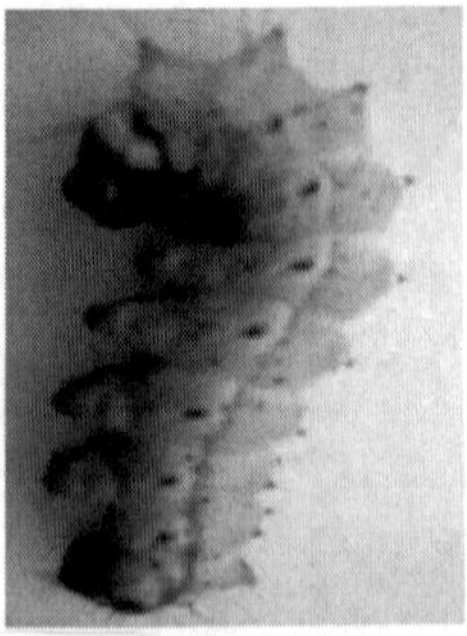

Early stage of infection: The infected muga silkworm larvae appear normal. Only microscopic examination of the silkworm larvae may indicate the presence of spore stage of the pathogen.

Later stage of infection:

- The silkworm larvae loose appetite, vary in size, retard in growth, moult irregularly and the colour of the larvae become light yellowish green instead of deep green colour of normal healthy larvae.
- Infected late stage larvae of muga silkworm show black dots or specks on the surface of the body and hence, the disease is known as 'Phutuka' i.e., spotted disease in Assamese.

Infection:

- The disease is transmitted from the infected mother to the offspring by transovarial/transovum means and this is called primary infection.
- If infection is primary, more than 50% larvae die before 3rdmoult and rarely any larva go for spinning.
- When healthy larvae get infected through contamination during rearing, it is called secondary infection.
- Secondary infection during early 4th larval stages leads to formation of flimsy cocoons.
- Whereas larvae infected during 5th larval stage form well-formed cocoons.

Infection source:

Egg stage

- Transovarially.
- Surface contamination of eggs (transovum).
- Contaminated grainage appliances.

Larval stage

- Contaminated egg laying kharika.
- Transovarially infected larvae.
- Faecal matters of infected larvae.
- Exuviae of infected larvae.
- Contaminated foliage.
- Contaminated rearing site.
- Contaminated rearing appliances.

Moth stage

- Purchase of infected seed cocoons
- Infected moth
- Infected grainage appliances.
- Meconium and moth scales.
- Grainage dust.

Spread of disease:

- Pebrinized larvae extrude faecal matter, gut juice and vomit containing pathogens, which contaminate the rearing environment, appliances and host plant foliage.
- Mostly, consumption of contaminated foliage/egg shell results in infection and spread of the disease.

Prevention/control of the disease:

- Follow the scientific inspection method of individual mother moth testing for detection of pebrine in egg production.
- Practice disinfection of grainage appliances before and after every grainage operation with 2% formalin.
- Ensure use of microscopically tested disease-free disinfected eggs only.
- Practice surface sterilization of the eggs with 2% formalin for 5 minutes.
- Maintain hygienic conditions in egg production room and rearing sites.
- For basic stock maintenance, follow cellular method of rearing.
- Practice disinfection of rearing appliances before use.
- During rearing, test the faecal matters, unequal/lethargic/unsettled/irregular moulters periodically. If pebrine spores are detected, reject the entire infected crop.
- Ensure the measures for destruction of diseased silkworm larvae/cocoons/moths/eggs.

Viral disease: Nuclear polyhedrosis, commonly known as Grasserie is a major viral disease of muga silkworm caused by a baculovirus.

Occurrence: The disease prevails althrough the year but is predominant during rainy summer months of the year.

Symptoms: The silkworm larva fails to moult. The integument becomes fragile and intersegmental region becomes swollen and that is why, the disease is known as 'Phularog' or swelling disease in Assamese. The body tissues and haemolymph of the infected larvae get disintegrated into turbid white fluid and the larvae hang upside down with the anal claspers after dying. The turbid fluid contains large number of hexagonal polyhedral bodies.

Infection: The silkworm larvae get infected on feeding of contaminated foliage of the host plants.

Infection source: Disintegrating diseased silkworm, its body fluid and contaminated rearing site and appliances.

Pre-disposing factor: High temperature clubbed with high humidity, poor quality host plant leaves.

Spread of disease: The diseased silkworm larva extrudes the pathogen along with oozing of body fluid due to injury and breakage of dead diseased larvae. The body fluid and broken body parts of the larvae contaminate the foliage, rearing site and appliances. The disease spreads to healthy worms on feeding of the contaminated leaves as well as the use of contaminated appliances during rearing.

Preventive measure of the disease:

- Practice disinfection of rearing site before rearing with 2% formalin solution spray. Dusting of 0.3% slaked lime in addition to usual disinfection procedure is recommended for rearing site and appliances disinfection in case of high incidence of disease in previous rearing.
- Pick out growth retarded/lethargic/irregular moulters and destroy.
- Ensure measures of destruction of diseased/doubtful worms by burning or burying with 5% formalin solution.
- Ensure hygienic conditions during rearing.
- Use certified disinfected disease free layings only.
- Ensure rearing on good quality leaves, because in case of silkworm, food is the major source of the diseases.

Bacterial disease

***Pseudomonas* bacterial disease.**

- This disease is abundantly found in muga plantation throughout the year.

Symptoms of disease

- Larva becomes lethargic and stop feeding
- Retarded growth of larva
- Vomit gut juice and excrete semi solid faeces,
- Sealing of anal lips
- Larva becomes soft and translucent.

Infectious flacherie

Symptoms:

- The infected silkworms are sluggish in movement.
- Lethargic and week.
- The anal portion of the larvae become swollen and ruptured.
- After 2 - 3 days of infection the silkworms die

Extent of damage:

- The disease is recorded during 4th and 5th inster silkworm in Aherua pre-peed crop.
- May-June with 12 - 22% crop loss. Management
- Feeding of quality and stage specific leaf.
- Dusting of rearing field with lime powder.
- Maintain hygienic condition of rearing field.

Other bacterial diseases

Chain type excreta, Sealing of anal lips and Rectal protrusion

Occurrence: These diseases prevails in all seasons of the year but is intensive during rainy summer months (June to August) of the year.

Pre-disposing factors

High temperature, high humidity, fluctuations in temperature and humidity conditions, inferior quality leaves, improper rearing, overcrowding of worms, frequent handling of worms.

Symptoms:

- The infected silkworm larvae become lethargic and motionless.
- The colour of the hemolymph turns black.
- Chain type excreta, sealing of anal lips, rectal protrusion are some of the easily detectable symptoms of the disease.
- Infected larvae die within a short time.

Infection: The muga silkworm larvae get infected on feeding of contaminated/poor quality foliage of the host plants.

Infection Source: Diseased larvae, its gut juice, faecal matters, body fluid and contaminated rearing site and appliances.

Pre-disposing factor: Sudden fluctuation in temperature and humidity, bad weather, poor quality leaves with high water content.

Spread of disease:

- The disease is transmitted by secondary infection of larvae feeding on the contaminated/poor quality leaves.
- Infected worms are oozing out body fluid containing pathogen throughout incubation period of infection and contaminate the leaves of the host plant and rearing environment.
- The disease spreads to healthy worms on feeding of the contaminated leaves. Feeding of late stage worms with very tender succulent leaves and sudden fluctuation of temperature and humidity during rearing period also may lead to outbreak of the disease.

Preventive measures of the disease:

- Use disinfected quality seeds of disease-free zone.
- Practice orientation of brushing to protect the young larvae from direct sunlight.
- Disinfection of rearing site before rearing with 2% formalin solution is mandatory. Ensure dusting of 0.3% slaked lime in addition to usual disinfection procedure recommended for rearing site and appliances in case of high incidence of the disease in previous rearing.
- Inspect rearing field regularly and pick out stunted/sluggish/irregular moulters and destroy.
- Ensure measures of destruction of diseased/doubtful worms by burying with 5% formalin solution.
- Practice washing of hands with formalin solution at the time of transfer of worms.

- Maintain hygienic conditions during rearing.
- Ensure rearing on good quality leaves, because in case of silkworm, food is the major vector of the diseases.
- Do not allow late stage worms to feed on tender succulent leaves.

Grassarie

- The disease is caused by virus.

Symptoms

- Infectious flacherie infected larvae
- The silkworm larvae fail to moult.
- The integument becomes fragil and inter segmental portions becomes swollen.
- Also stretched the skin.
- The body tissues and haemolymph of the infected larvae get disintegrated into turbid white fluid

Grassarie infected larvae

- The turbid fluid contains large number of hexagonal polyhedral bodies.

Source of infection

- Feeding of contaminated foliage.
- Disintegrating diseased silkworms, their body fluids.
- Contaminated rearing sites and appliances.

Predisposing factors

- High temperature and humidity with poor quality leaves are the major predisposing factors.

Management

- Disinfection of rearing site with 2% formalin solution before rearing.
- Dust 3% slaked lime in addition to usual disinfection in case of high incidence of disease in preceding rearing.
- Pick out growth retarded/ lethargic/ irregular moulters and destroy.
- Feeding of quality and stage specific leaf.
- Ensure proper hygiene during rearing. .

Mascardine

Causal Organism: *Beauvoria bassiana*

Mascardine is one of the major diseases of silkworm. However, it is less prevalent in muga silkworm and occurs

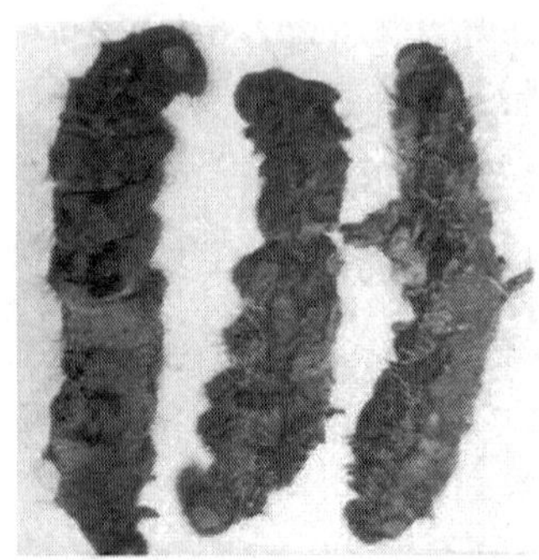

Symptoms

- The infected larvae loose the appetite .
- Become inactive.
- The colour of the larvae change to pale .
- Ceased the movement within 12-18 hours of infection.
- Within another 6 -8 hours the larvae die and gradually whole body of the larvae covered with white encrustation.

Source of infection

- Contaminated soil.
- Mummified / diseased larvae.

Mascardine infected larvae

- Contaminated rearing environment.

Spread of the disease

- The conidia/ spores of the pathogenic fungus are dispersed by wind.
- The conidia on contact with larval integument germinate, penetrate into the larval body and cause infection.

Extent of damage:

- Winter seasons, when night temperature fall down and day temperature high with foggy weather is the most favourable for out break of it.
- Upto 57- 88 % crop loss

Management

- Brushing of newly hatch worms in the sunny places during winter.
- Disinfect the rearing field with lime powder and bleaching powder before and after rearing.
- Dusting of slaked lime during rearing period to maintain the humidity.
- Maintain hygienic condition.

- Provide silk worm stage wise quality leaves.
- Use the "LAHDOI" developed by Pathology section ,CMER&TI.
- Collection and destruction of dead/ diseased larvae
- Pickout sick or dead worms with forceps/ chopstick and put in 2% formalin solution.
- Bury the carcasses in a pit and cover the soil.
- Wash hands with formalin or dettol solution after handling dead or infected larvae.
- Do not allow birds, ants or poultry to eat the carcasses.

Muscardine:

- It is a major disease of muga which mostly affects the pre seed crop.

Muga crop is divided into three crops

i. Commercial crop,

ii. Seed crop,

iii. Pre-seed crop.

Pre seed crops, Aherua (June –July and Jarua (December-January) is conducted during most adverse seasons and as such only 20-30 cocoons are harvested per dfl. 60- 90 per cent loss is encountered in Jarua crop.

Types

1. **Whitemuscardine** – *Beauveriabassiana*
2. **Green muscardine** – *Metarrhiziumanisopliae*

– Nomuraearileyi

Muscardine is one of the major diseases found in silkworm. But in muga silkworm, while muscardine is less prevalent occurring under certain specific environmental influence only.

Observations over the last decade revealed that the disease appeared alternatively after 2-3 years intervals. The causal organism of the disease is a fungus, which is not yet identified in case of muga silkworm.

Occurrence: The disease occurs during winter months of the year, when night temperature falls down and the day temperature remains comparatively high associated with high humidity i.e., foggy weather.

Symptoms: The infection may occur in any stage of the silkworm larvae. Infected larvae loose their appetite and become inactive. The colour of the larva turns pale, gradually ceases movement within 12-18 hours of infection.

The larvae hangs on tree twig or trunk and hardens. Within another 6-8 hours, the larva dies in next 16-18 hours, a white encrustation appears on the larval body. Within another 24 hours, the whole body of the larva gets covered by the white encrustation and becomes dry, brittle and mummifies.

Predisposing factors

- Low temperature with high humidity.

Source of infection: Mummified/diseased larvae contaminates the rearing environment and foliage of the host plants.

Spread of disease

- The conidia or spores of the pathogenic fungus are dispersed by wind and through contact. The conidia on contact with host integument, germinate, penetrate into the host body and cause the infection.
- Different chemical formulations viz., Labex, Resham Jyoti, Vijetha and Muga Guard as prophylactic measures were tested to control silkworm diseases.
- Muga Guard, a new formulation developed at the institute was tested for controlling bacterial and viral disease in muga silkworm.

Preventive measure of the disease

- Practice orientation of brushing towards sunlight during winter.
- Practice disinfection of rearing site before rearing with 2% formalin solution.
- Practice dusting of slaked lime in the field to control humidity at the time of rearing.
- Practice dusting of Tasar Kit Oushad developed by CTR&TI, Ranchi on the body of the larvae at the time of transfer.
- As prophylactic measure, spray 0.5% Sodium hydroxide solution on the worms after 24 hours of each moult.
- Maintain hygienic conditions during rearing.

Collection and destruction procedure of dead/diseased larvae

- Pick out sick or dead worms by forcep or chopstick and keep in a container with 2% formalin solution.
- Bury the carcasses in a pit and cover the soil.
- Wash hands with formalin or dettol solution after handling dead or infected larvae.
- Don't allow the birds, ants and poultry to eat the carcasses